Construction Mathematics

Construction Mathematics is an introductory level mathem[atics] written specifically for students of construction and relate[d]

- Learn by tackling exercises based on real-life constructio.. maths. Examples include: costing calculations, labour costs, cost of materials and setting out of building components.
- Suitable for beginners and easy to follow throughout.
- Learn the essential basic theory along with the practical necessities.

The second edition of this popular textbook is fully updated to match new curricula, and expanded to include even more learning exercises. End of chapter exercises cover a range of theoretical as well as practical problems commonly found in construction practice, and three detailed assignments based on practical tasks give students the opportunity to apply all the knowledge they have gained.

    Construction Mathematics addresses all the mathematical requirements of Level 2 construction NVQs from City & Guilds/CITB and Edexcel courses, including the BTEC First Diploma in Construction. Additional coverage of the core unit Construction Mathematics from BTEC National Construction, Civil Engineering and Building Services courses makes this an essential revision aid for students who do not have Level 2 mathematics experience before commencing their BTEC National studies. This is also the ideal primer for any reader who wishes to refresh their mathematics knowledge before going into a construction HNC or BSc.

Surinder Virdi is a lecturer in construction at South and City College Birmingham. He worked as a structural engineer for a number of years before starting his teaching career in further education. He has been teaching mathematics, construction science and construction technology on BTEC National and Higher National courses for the last 25 years.

Roy Baker spent forty years teaching mathematics, construction science and structural mechanics at the City of Wolverhampton College before retiring in 2005. He now works as a visiting lecturer.

Narinder Kaur Virdi is a programme manager at the South and City College Birmingham and responsible for managing the provision of basic skills. She worked as a biochemist for a number of years before becoming a trained teacher. She has been teaching literacy and numeracy for the last 15 years.

This is an excellent resource for construction students with step by step instructions helping them to understand and enjoy mathematics. Each chapter has an instruction and methodology section with solutions and testing at the end, enabling individual tracking and progression. This book will help my students improve their numeracy skills to achieve their diploma qualification.

Michael Cook, Lecturer, The Sheffield College

Construction Mathematics has proved to be an ideal text for Construction and Civil Engineering students at National Certificate level, whilst providing Undergraduates with a well laid out revision text. The logical progression of ideas and the simple and clear examples prove valuable to students as they pass onto a higher level of study.

Derek Spalton, Senior Lecturer, University of Derby

Construction Mathematics

Second Edition

Surinder Virdi, Roy Baker and Narinder Kaur Virdi

Routledge
Taylor & Francis Group

LONDON AND NEW YORK

First edition published 2007
by Butterworth Heinemann, an imprint of Elsevier

This edition published 2014
by Routledge
2 Park Square, Milton Park, Abingdon, Oxon OX14 4RN

and by Routledge
711 Third Avenue, New York, NY 10017

Routledge is an imprint of the Taylor & Francis Group, an informa business

British Library Cataloguing in Publication Data
A catalogue record for this book is available from the British Library

Library of Congress Cataloging-in-Publication Data
Virdi, Surinder Singh.
  Construction mathematics. — Second edition / Surinder Virdi, Roy Baker, and Narinder Kaur Virdi.
    pages cm
  Includes bibliographical references and index.
  1. Building — Mathematics.   I. Baker, Roy T.   II. Virdi, Narinder Kaur.   III. Title.
  TH437.V57 2014
  510.24'624 — dc23

                                                                          2012046094

ISBN13: 978-0-415-81078-4 (pbk)
ISBN13: 978-0-203-42780-4 (ebk)

Typeset in Helvetica
by Keystroke, Station Road, Codsall, Wolverhampton

MIX
Paper from
responsible sources
FSC
www.fsc.org   FSC® C013056

Printed and bound in Great Britain by
T International Ltd, Padstow, Cornwall

Contents

Preface to the second edition

This book is intended to provide the essential mathematics required by students on construction technician/craft courses. It covers the learning outcomes of the mathematics part of the unit 'Construction Science and Mathematics' for BTEC Level 2 Diploma course in construction. The book is also intended to help students studying the subject of Mathematics in Construction and Built Environment in the BTEC National/Extended Diploma in construction/civil engineering/building services and Higher National Certificate/Diploma courses in construction, although these syllabuses are not covered in their entirety.

Little previous knowledge is needed by students who use this text. The basic concepts and examples are explained in such a way that those construction students whose first interest is not mathematics will find it easy to follow. The contents have been divided into 20 chapters, providing information on a range of topics in algebra, geometry, trigonometry and statistics. Wherever applicable the basic concepts have been used to solve practical tasks in construction. Three assignments and 20 exercises are included to check and reinforce readers' learning.

This edition includes the solution of all questions included in the end of chapter exercises. The solutions will be beneficial to learners who may need some help in arriving at the right answers while solving questions from the exercises.

The authors would like to thank their students and colleagues for the interest they have shown in this project. A big thank-you to: Brian Guerin (Commissioning Editor, Routledge), Alice Aldous (Editorial Assistant) and Alanna Donaldson (Production Editor) for their support during the publication of this edition.

S.S. Virdi
R.T. Baker
N.K. Virdi

Preface to the first edition

This book is intended to provide the essential mathematics required by construction craft students. It covers the learning outcomes of the mathematics part of the unit Construction Science and Mathematics for BTEC First Diploma course in construction. The book is also intended to help construction students studying the subject of analytical methods in the BTEC National Diploma/Certificate in Construction and BTEC National Certificate in Civil Engineering, although these syllabuses are not covered in their entirety.

Little previous knowledge is needed by students who use this text. The basic concept and examples are explained in such a way that those construction students whose first interest is not mathematics will find it easy to follow. There are 20 exercises and two assignments for the students to check and reinforce their learning.

The authors would like to thank their wives, Narinder and Anne, for the encouragement and patience during the preparation of this book. The authors extend their thanks to the publishers and their editors, Rachel Hudson (Commissioning Editor) and Doris Funke, for their advice and guidance.

S.S. Virdi
R.T. Baker

How to use this book

Students pursuing level 3/4 courses should study all chapters in this book. Students pursuing level 2 courses should study all chapters except Chapters 4, 17, 18 and 19. The solutions for all questions included in the end of chapter exercises are given in Appendix 2. The reader should first try to solve the questions on their own, and only refer to the solutions if they experience some difficulty in finding the right answer.

Chapter 21 has three contextualised assignments, which the reader should attempt after studying topics relevant to the assignment tasks. The solutions are given online at www.routledge.com/9780415810784.

Acknowledgements

We are grateful to HMSO for permission to quote regulations on stairs from Building Regulations – Approved Document K.

Abbreviations and units

Units

Quantity	Name of unit	Symbol
Length	millimetre	mm
	centimetre	cm
	metre	m
Mass	kilogram	kg
Time	second	s
Fahrenheit temperature	degrees Fahrenheit	°F
Plane angle	radian	rad
	degrees	°
Force	Newton	N
Stress	stress	N/mm^2
		kN/mm^2
Fluid pressure	Pascal	Pa
Acceleration due to gravity	g	m/s^2
Potential difference	volts	V
Electrical resistance	Ohms	Ω
Celsius temperature	degrees Celsius	°C
Coefficient of thermal transmittance	U-value	W/m^2 °C
Heat loss	watt	W
Luminous intensity	candela	cd
Illuminance	lux	lx

Prefixes

Name	Symbol	Factor
Tera	T	10^{12}
Giga	G	10^{9}
Mega	M	10^{6}
Kilo	k	10^{3}
Hecto	h	10^{2}
Deca	da	10
Deci	d	10^{-1}
Centi	c	10^{-2}
Milli	m	10^{-3}
Micro	μ	10^{-6}
Nano	n	10^{-9}
Pico	p	10^{-12}

Abbreviations/symbols

Description	Symbol
Approximately	Approx.
For example	e.g.
Triangle	Δ
Less than	<
Greater than	>
Degrees, minutes, seconds	° ' "
Angle	∠
Significant figures	s.f.
Therefore	∴
Decimal places	d.p.

CHAPTER 1

Using a scientific calculator

Learning outcomes:

(a) Identify the right keys to perform a calculation

(b) Perform a range of calculations

1.1 Introduction

The use of electronic calculators became popular during the early 1970s. Before the invention of calculators, slide-rules and tables of logarithms and antilogarithms were used to perform simple as well as complex calculations. The exercises and assignments in this book require the use of a scientific calculator; this chapter deals with the familiarisation of some of the main keys of a calculator.

With most calculators the procedure for performing general calculations is similar. However, with complex calculations this may not be the case. In that situation the reader should consult the instructions manual that came with their calculator. The dissimilarity in calculators is not just limited to the procedure for calculations, as the layout of the keys could be different as well.

The sequence in which the keys of a new calculator are pressed is the same as the sequence in which a calculation is written. With the old calculators this might not be the case. All calculations given in this section are based on the new calculators. Scientific calculators have a range of special function keys and it is important to choose one that has all the functions most likely to be needed. Some of the commonly used keys are shown in Section 1.2.

1.2 Keys of a scientific calculator

The keys of a typical scientific calculator are shown in Figure 1.1.

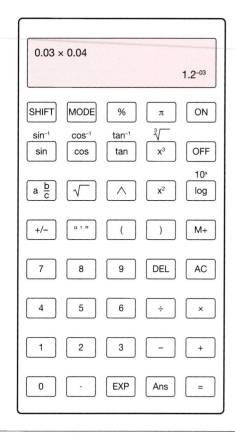

Figure 1.1

Key	Description
+	Adds two or more numbers
−	Subtracts a number from another
÷	Divides a number by another
×	Used to multiply two or more numbers
AC	Cancels or clears an existing calculation
SHIFT	Press this key to use the second function of a key
MODE	Use this key to set the calculator for performing calculations in terms of degrees or radians
√	Calculates the square root of a number
$\sqrt[3]{}$	Calculates the cube root of a number
x^2	Use this key to determine the square of a number
x^3	Use this key to determine the cube of a number
^ or $x^\square$	A number can be raised to any power by pressing this key
π	Use this key wherever π occurs in a formula
sin cos tan	Use the appropriate key to determine the sine/cosine/tangent of an angle

sin⁻¹ cos⁻¹ tan⁻¹	If the sin/cos/tan of an angle is given, use the appropriate key to determine the angle
log	Use this key if the calculation involves logarithm to the base 10
10^x	This key is used to calculate antilogarithms, i.e. the reverse of log
EXP	Use this key to raise 10 to the power of a given number
a$\frac{b}{c}$	Use this key to perform calculations involving fractions
M⁺	This key is used to input values into memory
%	Press this key to express the answer as a percentage
°'"	This key is used to convert an angle into degrees, minutes and seconds
( )	These keys will insert brackets in the calculations involving complicated formulae.
DEL	Press this key to delete the number at the current cursor position

Example 1.1

Calculate 37.80 – 40.12 + 31.55

Solution:

The sequence of pressing the calculator's keys is:

3	7	.	8	0	–	4	0	.	1	2	+

3	1	.	5	5	=	**29.23**

Example 1.2

Calculate $\dfrac{34.9 \times 57.3}{41.66}$

Solution:

The sequence of inputting the information into your calculator is:

3	4	.	9	×	5	7	.	3	÷	4	1	.	6

6	=	**48.0**

Example 1.3

Calculate $\dfrac{87.3 \times 67.81}{23.97 \times 40.5}$

Solution:

This question can be solved in two ways. The calculator operations are:

(1) 87.3 × 67.81 ÷ 23.97 ÷ 40.5

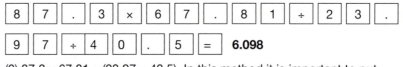

(2) 87.3 × 67.81 ÷ (23.97 × 40.5). In this method it is important to put 23.97 × 40.5 within brackets. Failure to do so will result in the wrong answer.

Example 1.4

Calculate $\sqrt{4.5} \times \sqrt{5.5} + \sqrt{3.4}$

Solution:

The calculator operation is shown below:

Example 1.5

Calculate the value of πr^2 if $r = 2.25$

Solution:

The calculator operation is:

Example 1.6

Find the value of (2.2 × 4.8) + (5.2 × 3)

Solution:

The sequence of calculator operation is:

Example 1.7

Evaluate $\dfrac{6^3 \times 4^4}{2^5}$

Solution:

In this question the $\boxed{\wedge}$ key will be used to raise a number to any power.
Press the following keys in the same sequence as shown:

$\boxed{6}$ $\boxed{\wedge}$ $\boxed{3}$ $\boxed{\times}$ $\boxed{4}$ $\boxed{\wedge}$ $\boxed{4}$ $\boxed{\div}$ $\boxed{2}$ $\boxed{\wedge}$ $\boxed{5}$ $\boxed{=}$ **1728**

Example 1.8

Calculate $10 \log_{10}\left(\dfrac{4 \times 10^{-7}}{2 \times 10^{-12}}\right)$

Solution:

The $\boxed{\text{EXP}}$ key will be used to raise 10 to any power, as shown:

$\boxed{1}$ $\boxed{0}$ $\boxed{\text{LOG}}$ $\boxed{(}$ $\boxed{4}$ $\boxed{\text{EXP}}$ $\boxed{+/-}$ $\boxed{7}$ $\boxed{\div}$ $\boxed{2}$ $\boxed{\text{EXP}}$ $\boxed{+/-}$

$\boxed{1}$ $\boxed{2}$ $\boxed{)}$ $\boxed{=}$ **53.01**

Example 1.9

Calculate $\dfrac{\sin 60°}{\cos 60°}$

Solution:

The calculator must show D in the display area. If the calculator displays R or G then use the $\boxed{\text{MODE}}$ key or the $\boxed{\text{SETUP}}$ key to change the angle unit to degrees, and press the following keys:

$\boxed{\sin}$ $\boxed{6}$ $\boxed{0}$ $\boxed{\div}$ $\boxed{\cos}$ $\boxed{6}$ $\boxed{0}$ $\boxed{=}$ **1.732**

Example 1.10

Find the angle if: (a) the sine of an angle is 0.6
(b) the cosine of an angle is 0.45
(c) the tangent of an angle is 0.36

Solution:

Use the $\boxed{\text{MODE}}$ key or the $\boxed{\text{SETUP}}$ key to change the angle unit to degrees. As this question involves the determination of angles, the process is the reverse of that used in Example 1.9. Instead of sin, cos or tan keys, use $\sin^{-1}$, $\cos^{-1}$ and $\tan^{-1}$.

(a) Use the following sequence to determine the angle as a decimal number first, and then change to the sexagesimal system (i.e. degrees, minutes and seconds)

$\boxed{\text{SHIFT}}$ $\boxed{\sin}$ $\boxed{0}$ $\boxed{.}$ $\boxed{6}$ $\boxed{=}$ 36.8699° $\boxed{°\,'\,''}$ **36°52′11.6″**

(b) $\boxed{\text{SHIFT}}$ $\boxed{\cos}$ $\boxed{0}$ $\boxed{.}$ $\boxed{4}$ $\boxed{5}$ $\boxed{=}$ 63.2563° $\boxed{°\,'\,''}$ **63°15′22.7″**

(c) $\boxed{\text{SHIFT}}$ $\boxed{\tan}$ $\boxed{0}$ $\boxed{.}$ $\boxed{3}$ $\boxed{6}$ $\boxed{=}$ 19.7989° $\boxed{°\,'\,''}$ **19°47′56″**

Exercise 1.1

The solutions to Exercise 1.1 can be found in Appendix 2.

1. Calculate $37.85 - 40.62 + 31.85 - 9.67$.

2. Calculate $\dfrac{33.9 \times 56.3}{45.66}$.

3. Calculate $\dfrac{67.3 \times 69.81}{25.97 \times 20.5}$.

4. Calculate $\sqrt{4.9} \times \sqrt{8.5} + \sqrt{7.4}$.

5. Calculate the value of πr^2 if $r = 12.25$.

6. Find the value of: (a) $(2.2 \times 9.8) + (5.2 \times 6.3)$
   (b) $(4.66 \times 12.8) - (7.5 \times 5.95)$
   (c) $(4.6 \times 10.8) \div (7.3 \times 5.5)$.

7. Evaluate: (a) $\dfrac{5^3 \times 3^4}{2^5}$.

   (b) $\dfrac{4^3 \times 6^3}{5^4}$.

8. Calculate $10 \log_{10} \left(\dfrac{9 \times 10^{-8}}{2 \times 10^{-11}} \right)$.

9. Calculate: (a) $\dfrac{\sin 70°}{\cos 60°}$

   (b) $\dfrac{\tan 45°}{\cos 35°}$.

10. Find the angle if: (a) the sine of an angle is 0.85
    (b) the cosine of an angle is 0.75
    (c) the tangent of an angle is 0.66.

11. Calculate the values of: (a) $\sin 62°42'35''$
    (b) $\cos 32°22'35''$
    (c) $\tan 85°10'20''$.

Answers to Exercise 1.1

1. 19.41
2. 41.8
3. 8.83
4. 9.17
5. 471.44

6. (a) 54.32   (b) 15.02   (c) 1.24

7. (a) 316.41   (b) 22.12

8. 36.53

9. (a) 1.879   (b) 1.221

10. (a) 58°12′42″   (b) 41°24′34.64″   (c) 33°25′29.32″

11. (a) 0.8887   (b) 0.8445   (c) 11.8398

CHAPTER **2**

Numbers

Learning outcomes:

(a) Identify positive numbers, negative numbers, integers and decimal numbers

(b) Perform calculations involving addition, subtraction, multiplication and division

(c) Use order of operations (BODMAS) to perform calculations

2.1 Introduction

Mathematics involves the use of numbers in all of its branches like algebra, geometry, statistics, mechanics and calculus. The use of numbers also extends to other subjects like estimating, surveying, construction science and structural mechanics. As we shall be dealing with numbers in all sections of this book, it is appropriate to deal with the different types of numbers at this stage.

2.2 History of numbers

In early civilisations different types of counting systems were used in business and other fields. It all started with the use of lines, which later developed into alphabets (Rome, Greece), symbols (Babylon), hieroglyphics (Egypt), pictorials (China) and lines and symbols (India). The Roman numerals (I, V, X, L, C, D and M), although widely used in commerce and architecture had two major flaws. First, there was no zero; and second, for large numbers different types of systems were used.

Indian-Arab numerals, the forebears of modern numbers, were used in India more than 2500 years ago. Originally there were nine symbols to represent 1–9 and special symbols were used for tens, hundreds and thousands. It appears that the Indians later introduced zero in the form of a dot (to represent nothing), which they either borrowed from other systems or invented themselves. The credit for disseminating to the European countries goes to the Arabs, who started to expand their trade about 1500 years ago and had links with several countries. After some resistance, the use of Indian-Arab numerals became widespread during the sixteenth century and the Roman numerals were restricted to special use.

The main reasons for the universal adoption of the Indian-Arab numerals were that it was a place-value system and that it was also very easy to use. For example, if you were to write, say, 786 and 1998 in Roman numerals, just imagine the time taken to do so. In Roman numerals 786 is written as DCCLXXXVI. Similarly, 1998 is written as MCMXCVIII.

2.3 Positive numbers, negative numbers and integers

Numbers with either a plus (+) sign or no sign on their left are called positive numbers. For example:

2, +3, 5, 11, 5000

The greater a positive number, the greater is its value. Negative numbers have a minus (–) sign on their left. For example:

–3, –2, –21, –250.

The greater a negative number, the smaller its value. For instance, the value of –15 is less than –10. This can be explained by considering a thermometer, as shown in Figure 2.1.

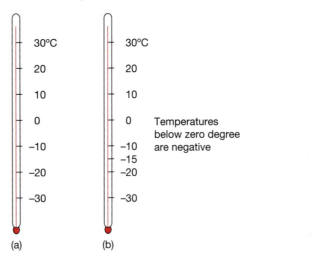

Figure 2.1

If the object is at –10°C, and its temperature is reduced by 5°C, the new temperature will be:

–10 – 5 = –15°C

What we see from Figure 2.1 can be generalised to say that the value of positive numbers is greater than zero, whereas the value of negative numbers is less than zero.

An integer is a whole number, positive or negative. For example, the following are integers:

10, –30, 0, 24, –270

2.4 Prime and composite numbers

Any positive integer having no factors other than itself and unity is called a prime number. Also, prime numbers are greater than 1. Some of the prime numbers are: 2, 3, 5, 7, 11, 13, 17, 19, 23, 29, 31, 37. For example, the factors of 7 are 7 and 1, as $7 \times 1 = 7$; similarly, the factors of 11 are 11 and 1, as $11 \times 1 = 11$.

Prime numbers can be divided only by their factors. For example:

$$\frac{7}{1} = 7 \text{ and } \frac{7}{7} = 1$$

A composite number can have other factors in addition to itself and 1. For example, 12, which is not a prime number, has the following factors:

$2 \times 6 = 12$

$2 \times 2 \times 3 = 12$

$3 \times 4 = 12$

$1 \times 12 = 12$

2.5 Square numbers

A number that can be obtained by squaring another number is called a square number; 1, 4, 9, 16, 25, 36, 49, 64, 81, 100 are some of the square numbers as they can be obtained by squaring 1, 2, 3, 4, 5, 6, 7, 8, 9 and 10, respectively.

2.6 Addition and subtraction

Addition involves combining two or more numbers to give the sum total. The plus sign (+) is used to denote addition. Subtraction involves taking a number away from another number. The minus or negative sign (–) is used to denote subtraction. The following rules apply to addition/subtraction:

(1) $x + y = x + y$ for example: $11 + 5 = 16$

(2) $x - y = x - y$ $11 - 5 = 6$

(3) $x + (-y) = x - y$ $11 + (-5) = 11 - 5 = 6$

(4) $x - (-y) = x + y$ $11 - (-5) = 11 + 5 = 16$

(5) $-x - y = -(x + y)$ $-11 - 5 = -(11 + 5) = -16$

Example 2.1

A bricklayer bought the following items at a local DIY store:

- Concrete bricks: £120.00
- Cement: £11.70
- Sand: £10.50

If the cost of delivering the materials was £12.00, find:

(a) The total amount of the bill

(b) The change he received if he gave £160 to the cashier.

Solution:

(a) Total amount of the bill = £120.00 + £11.70 + £10.50 + £12.00
= **£154.20**

(b) Change = £160.00 − £154.20 = **£5.80**

Example 2.2

The air temperature today at 9 p.m. was −1°C and is expected to be 6 degrees lower at 6 a.m. tomorrow. It is also expected that at 11 a.m. tomorrow the temperature will be 5 degrees higher than that expected at 6 a.m. Find the temperatures expected tomorrow at 6 a.m. and 11 a.m.

Solution:

The fall in temperature involves subtraction and the rise in temperature involves addition.

Temperature at 6 a.m. tomorrow = −1 − 6 = **−7°C**

Temperature at 11 a.m. tomorrow = −7 + 5 = **−2°C**

Example 2.3

Calculate the length of skirting board required for the room shown in Figure 2.2.

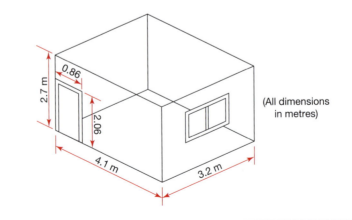

(All dimensions in metres)

Figure 2.2

Solution:

Length of skirting board = Total length of the walls – door width

$$= 4.1 + 3.2 + 4.1 + 3.2 - 0.86$$

$$= 14.6 - 0.86 = \textbf{13.74 m}$$

2.7 Decimal numbers

Decimals may be used when a number:

- is not divisible by another number, e.g. $106 \div 4 = 26.5$
- is less than 1 (or the numerator is less than the denominator). For example:

(1) $\dfrac{1}{10} = 0.1$

(2) $\dfrac{1}{25} = 0.04$

(3) $\dfrac{1}{100} = 0.01$

The number of digits after the decimal depends on the value of the denominator. In $\dfrac{1}{10}$, the denominator has only one zero. Therefore, the decimal should be put in the numerator after one digit, moving from right to left:

$$\dfrac{1}{10} = .1$$

In $\dfrac{1}{100}$ there are two zeros in the denominator. In this case the decimal is placed in the numerator after two digits. Since there is only one digit in the numerator, another digit needs to be added. This extra digit has to be a zero:

$$\dfrac{1}{100} = .01, \text{ the decimal put after two digits.}$$

In $\dfrac{1}{25}$, there are no zeros in the denominator, but by multiplying the numerator and denominator by 4, we have $\dfrac{1}{25} = \dfrac{1 \times 4}{25 \times 4} = \dfrac{4}{100} = .04$, with the decimal put after two digits.

If a number is smaller than 1, for example .04 or .1, it is standard practice to put a zero before the decimal, so .04 becomes 0.04 and .1 becomes 0.1.

2.7.1 Place value

In a number, each place has its own value. Some of the place values are:

thousands	for example	1000
hundreds		100
tens		10
units		1
tenths		$\frac{1}{10}$ or 0.1
hundredths		$\frac{1}{100}$ or 0.01
thousandths		$\frac{1}{1000}$ or 0.001

The decimal point is introduced between units and tenths.

Take a number, say 21.367. It is made up of 2 tens, 1 unit, 3 tenths, 6 hundredths and 7 thousandths, which give 21.367 when added:

2×10	1×1	3×0.1	6×0.01	7×0.001
$= 20$	$= 1$	$= 0.3$	$= 0.06$	$= 0.007$

$$20 + 1 + 0.3 + 0.06 + 0.007 = 21.367$$

A feature of decimals is that the addition of more zeros between a number and the decimal point reduces its value. Referring to the example, 0.06 is smaller than 0.3; similarly 0.007 is smaller than 0.06.

2.7.2 Adding, subtracting and multiplying decimals

The methods of carrying out the addition and subtraction of decimal numbers are similar to those used for whole numbers. However, because of the decimal, the numbers are written with their decimals aligned, as shown in Example 2.4.

Example 2.4

(a) Add 25.12, 106.239 and 8340.0191

(b) Subtract 237.347 from 645.591

Solution:

(a) Write the numbers with their decimals aligned and move from right to left:

$$
\begin{array}{r}
25.12 \\
+\ 106.239 \\
+\ \underline{8340.0191} \\
\mathbf{8471.3781}
\end{array}
$$

(b) Again, write the numbers with their decimals aligned, and proceed from right to left:

$$645.591$$
$$\underline{-237.347}$$
$$\mathbf{408.244}$$

In multiplication the easiest way is to ignore the decimals and multiply as done with whole numbers. The decimal point is inserted afterwards, as shown in Example 2.5

Example 2.5

Multiply 13.92 and 5.4

Solution:

Consider the given numbers to be whole numbers and multiply:

$$\begin{array}{r} 1392 \\ \times \underline{54} \\ 5568 \\ \underline{69600} \\ 75\ 168 \end{array}$$

Count the number of digits to the right of the decimal points in the numbers being multiplied:

In 13.92 there are two digits to the right of the decimal; in 5.4 there is only one digit to the right of the decimal. Therefore, the total number of digits to the right of the decimal in the two numbers is the sum of 2 and 1, i.e. 3.

Counting from right to left, put the decimal after three digits. Hence the answer is **75.168**.

2.7.3 Multiplication and division by the powers of 10

When a number (say x) is divided by another number that is greater than 1, the value of the answer will be less than x. The opposite is true when a number is multiplied by a number greater than 1. The process becomes very simple when a number is multiplied or divided by the powers of 10. This is illustrated in Examples 2.6 and 2.7

Example 2.6

Multiply   (a) 451 by 100

(b) 28.67 by 10

(c) 28.67 by 100

(d) 28.67 by 1000

Solution:

(a) An integer (whole number) becomes bigger in value when it is multiplied by a number greater than 1. When multiplying an integer by 10, 100, 1000, etc. simply add the appropriate number of zeros to the original number. The multiplier in this question (i.e. 100) has two zeros, therefore put two zeros after writing 451 to make the answer bigger:

$451 \times 100 =$ **45 100**

(b) A decimal number also becomes bigger on multiplying with a number greater than 1. To do this, move the decimal point to the right by the same number of places as there are zeros in the multiplier. As there is only one zero in 10, move the decimal point by one place to a position between 6 and 7:

$28.67 \times 10 =$ **286.7**

(c) As 100 has two zeros, move the decimal to the right by two places:

$28.67 \times 100 = 2867.$

As there is nothing after the decimal, $2867. =$ **2867**

(d) Following the procedure explained in (b) and (c) move the decimal point by three places. Add zero or zeros to fill the blank space.

$28.67 \times 1000 = 2867\_. = 28\ 670. =$ **28 670**

Example 2.7

Divide (a) 451 by 100

        (b) 28.67 by 100

        (c) 28.67 by 1000

Solution:

Division will give an answer that is smaller in value than the original number. This will involve introduction/movement of the decimal point, moving from right to left. The number of places the decimal point has to be moved will be equal to the number of zeros in the divisor.

(a) The divisor (i.e. 100) has two zeros. Place the decimal after two digits, moving from right to left:

$$\frac{451}{100} = \textbf{4.51}$$

(b) The solution to this question will involve the movement of the decimal point by two places to the left:

$$\frac{28.67}{100} = .2867 = \textbf{0.2867}$$

(c) To solve this question, move the decimal point by three places to the left. This will leave a blank space, which should be filled with a zero.

$$\frac{28.67}{1000} = ._2867 = .02867 = \mathbf{0.02867}$$

2.8 Order of operations

A calculation in arithmetic and algebra may involve one or more of: addition, subtraction, multiplication and division. If we are asked to evaluate 3 + 4 × 2, our answer could be 11 or 14:

3 + 4 × 2 = 3 + 8 = 11

3 + 4 × 2 = 7 × 2 = 14

For calculations of this nature an order of precedence of operation, or in other words the order in which the calculations are to be done, is used. This can be remembered using the acronym BODMAS, which stands for:

B   brackets

O   of (same as multiplication, e.g. ½ of 8 = ½ × 8 = 4)

D   division

M   multiplication

A   addition

S   subtraction

The correct evaluation of 3 + 4 × 2 is 11 as the multiplication is done before addition. Modern calculators will automatically carry out all calculations using the rules of precedence.

2.8.1 Brackets

Brackets are used in mathematical problems to indicate that the operation inside the brackets must be done before the other calculations. For example:

(4 − 2) × 10 = 2 × 10 = 20

There are many ways in which we can have brackets in mathematical calculations. Some of the rules that we need to follow are:

- If there is a number in front of a bracket, then multiply the contents of the brackets by that number.
- If there is no number before or after a bracket, then assume the number 1 and multiply the contents of the brackets by 1.

Example 2.8

Solve: (a) 4 − 2 + 3(4 × 1.5) + 5

(b) 12 − (2 × 4 + 2)

Solution:

(a) $4 - 2 + 3(4 \times 1.5) + 5 = 4 - 2 + 3(6) + 5 = 4 - 2 + 18 + 5 = \mathbf{25}$

(b) $12 - (2 \times 4 + 2) = 12 - \mathbf{1}(8 + 2) = 12 - \mathbf{1}(10) = 12 - 10 = \mathbf{2}$

Exercise 2.1

The solutions to Exercise 2.1 can be found in Appendix 2.

1. Add 34.21, 26.05 and 370.30 without using a calculator.
2. Subtract 20.78 from 34.11 without using a calculator.
3. Solve $309.1 - 206.99 - 57.78$.
4. Multiply 40.0 and 0.25 without using a calculator.
5. Solve $25 \times 125 \div 625$.
6. Multiply: (a) 379 by 100
                 (b) 39.65 by 1000
                 (c) 39.65 by 10 000.
7. Divide: (a) 584 by 100
                 (b) 45.63 by 100
                 (c) 45.63 by 1000.
8. Solve: (a) $6 - 2 - 3(4 \times 1.5 \times 0.5) + 15$
                 (b) $15 - (2 \times 4 + 2) - (4 - 3)$.
9. The minimum temperature on Monday was –2°C and by 0600 hours on the next day the temperature had dropped by a further 5°C. Find the temperature at 0600 hours on Tuesday.
10. Bob Sands bought the following items at a DIY store (all prices include VAT): emulsion paint: £29.85; non-drip gloss paint: £19.98; paint brushes: £15.50.

    (a) Find the total amount of the bill.

    (b) If Bob gave the cashier four £20 notes, find the amount of change he received.
11. Figure 2.3 shows Amanda's dining room, where she wants to replace the old skirting board and the coving.

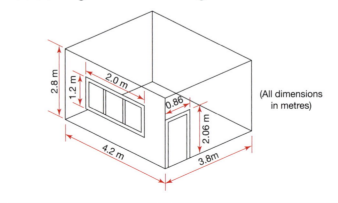

(All dimensions in metres)

Figure 2.3

Find: (a) the cost of buying the skirting boards if they are available in 2.4 m lengths, and a pack of four costs £34.99

(b) the cost of buying the coving, if they are available in 3.0 m lengths, and cost £6.00 per length.

12. Find the cost of changing the wallpaper in the room shown in Figure 2.3 if each roll of the selected wallpaper costs £21.99. Include a lump sum of £25.00 for the wallpaper glue and other accessories. Each roll of wallpaper consists of a 52 cm wide and 10.0 m long sheet; assume the wastage of wallpaper is 15 per cent. The size of the window is 2.0 m × 1.2 m high, and the width of the skirting board as well as the coving is 100 mm.

Answers to Exercise 2.1

1. 430.56
2. 13.33
3. 44.33
4. 10.0
5. 5
6. (a) 37 900   (b) 39 650   (c) 396 500
7. (a) 5.84   (b) 0.4563   (c) 0.04563
8. (a) 10   (b) 4
9. –7°C
10. (a) £65.33   (b) £14.67
11. (a) £69.98   (b) £36.00
12. £222.91

CHAPTER 3

Basic algebra

Learning outcomes:

(a) Add and subtract algebraic expressions

(b) Multiply and divide algebraic expressions

(c) Solve linear equations

3.1 Introduction

In algebra numbers are used with letters, the latter denoting anything like age, area, volume, temperature, etc. Using algebraic notation we can form equations that can be used to solve many mathematical problems. The rules of adding, subtracting, multiplying and dividing algebraic expressions are identical to those used for numbers. In algebraic expressions, however, we must distinguish between the different expressions. For example, in $x^2 - 4x + 5$, x^2 and $4x$ are different expressions and cannot be subtracted as such.

3.2 Addition and subtraction

Only similar expressions can be added or subtracted as the addition and subtraction of dissimilar expressions will result in a wrong answer. In the expression $2a + 3b - a + b$, symbols a and b may be assumed to represent aerated concrete blocks and bricks, respectively. As the two materials are different, the terms with 'a' ($2a$ and $-a$) need to be processed separately from the terms with 'b' – i.e. $3b$ and b.

$$2a + 3b - a + b = 2a - a + 3b + b$$
$$= 2a - 1a + 3b + 1b \ (a = 1a; b = 1b)$$
$$= a + 4b \ (\text{simplify } 2a - 1a, \text{ separately from } 3b + 1b)$$

Example 3.1

Simplify (a) $5a + 2b - 3b + 2a$

(b) $-5x - 3y + 2x - 3y$

Solution:

(a) Rearrange the terms of $5a + 2b - 3b + 2a$:

$$5a + 2b - 3b + 2a = 5a + 2a - 3b + 2b$$
$$= \mathbf{7a - b} \quad (5a + 2a = 7a; -3b + 2b = -1b \text{ or } -b)$$

(b) Rearrange the terms of $-5x - 3y + 2x - 3y$:

$$-5x - 3y + 2x - 3y = 2x - 5x - 3y - 3y$$
$$= \mathbf{-3x - 6y} \quad (2x - 5x = -3x; -3y - 3y = -6y)$$

3.3 Multiplication and division

Multiplication of letters is done in the same manner as the multiplication of numbers. For example:

$$5 \times 5 = 5^2$$

$a \times a = a^2$ (also see Chapter 4 on indices)

As in addition and subtraction, only the similar terms are multiplied. However, in addition and subtraction y^2 and y^3 are treated as different terms, but in multiplication and division they can be multiplied and/or divided as the base – in this case 'y' – is the same. Examples 3.2 and 3.3 explain the processes of multiplication and division involving simple algebraic expressions. For complex questions (for example: multiply $5a^2 - 2a + 6$ by $2a^2 + 8a - 3$) refer to a textbook on analytical methods.

Example 3.2

Simplify $2xy^2 \times 5x^2y^3$

Solution:

As the numbers, x terms and y terms are different from one another, they will be multiplied separately first and then combined to give the answer.

$$2 \times 5 = 10 \tag{1}$$
$$x \times x^2 = x \times x \times x = x^3 \tag{2}$$
$$y^2 \times y^3 = y \times y \times y \times y \times y = y^5 \tag{3}$$

Combining (1), (2) and (3):

$$2xy^2 \times 5x^2y^3 = \mathbf{10\ x^3y^5}$$

Example 3.3

Divide $25a^3\,b^2c^4$ by $5a^2\,bc^2$

Solution:

Mathematically the question can be written as: $25a^3b^2c^4 \div 5a^2bc^2$, or $\dfrac{25a^3b^2c^4}{5a^2bc^2}$. The numbers and other terms will be simplified individually first, but combined later to give the answer:

$$\frac{25}{5} = 5 \qquad\qquad\qquad (1)$$

$$\frac{a^3}{a^2} = \frac{a \times a \times a}{a \times a} = a \qquad\qquad\qquad (2)$$

$$\frac{b^2}{b} = \frac{b \times b}{b} = b \qquad\qquad\qquad (3)$$

$$\frac{c^4}{c^2} = \frac{c \times c \times c \times c}{c \times c} = c \times c = c^2 \qquad\qquad\qquad (4)$$

Combining 1, 2, 3 and 4:

$$\frac{25a^3b^2c^4}{5a^2bc^2} = \mathbf{5abc^2}$$

Example 3.4

Multiply $2x + 3$ by $x - 1$

Solution:

Write the two expressions as shown:

$$2x + 3$$
$$\times \underline{x - 1}$$

Multiplication in algebra proceeds from left to right. $2x + 3$ is multiplied by x first and then by -1. After multiplication the terms are added or subtracted, if necessary:

$$
\begin{array}{r}
2x + 3 \\
\times\; \underline{x - 1} \\
2x^2 + 3x \\
\underline{-\, 2x - 3} \\
\mathbf{2x^2 +\; x - 3}
\end{array}
$$

3.4 Brackets

Brackets may be used in algebra to simplify expressions, taking the common factors out and leaving the rest within. In $2x + 6y$, 2 is a common factor as 2 and 6 are both divisible by 2.

Therefore, $2x + 6y$ can be written as $2(x + 3y)$. As there is no mathematical sign between 2 and the bracket, it means that both x and $3y$ will be multiplied by 2 when the brackets are removed:

$$2(x + 3y) = 2 \times x + 2 \times 3y = 2x + 6y$$

In an expression like $2 + (3x - 4y)$, there is no number or symbol just before the brackets. There is a $+$ sign, which implies that nothing will change if the brackets are removed,

$$2 + (3x - 4y) = 2 + 3x - 4y$$

A minus sign in front of the brackets would mean changing the signs of the terms within the brackets, if simplification is required,

$$2 - (3x - 4y) = 2 - 3x + 4y \quad (3x \text{ becomes } -3x \text{ and } -4y \text{ becomes } +4y)$$

Example 3.5

Simplify: (a) $3(2x - 3y)$

(b) $2(-4x + 2y)$

(c) $6 + (x + 2y + 3z)$

(d) $6 - (x + 2y - 3z)$

Solution:

(a) Simplify $3(2x - 3y)$ by removing the brackets. As explained earlier, the removal of brackets will involve the multiplication of $2x$ and $-3y$ by 3:

$$3(2x - 3y) = 3 \times 2x - 3 \times 3y$$
$$= \mathbf{6x - 9y}$$

(b)

$$2(-4x + 2y) = 2 \times -4x + 2 \times 2y$$
$$= \mathbf{-8x + 4y}$$

(c) In this question there is no symbol or number, but a plus sign in front of the brackets. The signs of the terms that are within the brackets will not change if the removal of brackets is undertaken:

$$6 + (x + 2y + 3z) = \mathbf{6 + x + 2y + 3z}$$

(d) In this question the signs of the terms within the brackets will change, as there is a minus sign before the brackets:

$$6 - (x + 2y - 3z) = \mathbf{6 - x - 2y + 3z}$$

3.5 Simple equations

An equation is a mathematical statement that shows the equality of two expressions. The two expressions are separated by the $=$ sign. There are several types of equation, but an equation which has only one symbol and in which the symbol is only raised to the power 1, is known as a simple equation. A simple equation is also known as a linear equation as it can be represented graphically by a straight line.

Consider the equation $x + 4 = 6$. The left hand side (LHS) of this equation, $x + 4$, must be equal to 6, the right hand side (RHS) of the equation. The value of the unknown, x, should be such that the LHS is also equal to 6. In other words x must be equal to 2. Finding the value of the unknown quantity in an equation is known as 'solving the equation'. The solution of simple equations may require:

- addition or subtraction of symbols or numbers
- multiplication or division by symbols or numbers.

It is important to remember that whatever is done to one side of the equation, the same must be done to the other side to maintain the equality of the two sides. Example 3.6 illustrates this point.

Example 3.6

Solve (a) $x - 6 = 12$

(b) $3a = 15$

(c) $\dfrac{y}{5} = 3$

(d) $2x - 7 - 3x = 3 - 5x$

Solution:

(a) *Method 1*: In questions of this type we can either add or subtract a number to leave x, the unknown quantity, on its own. The same operation should be carried out on the other side of the equation as well. In this example, 6 is added to both sides of the equation, which in turn leaves the LHS simplified:

$x - 6 + 6 = 12 + 6$

or **$x = 18$**

Method 2: Transfer –6 to the RHS. It becomes +6 in this process

$x - 6 = 12$

$x = 12 + 6 = $ **18**

(b) To solve this question, both sides will be divided by 3 to leave a on its own:

$\dfrac{3a}{3} = \dfrac{15}{3}$

$a = 5$ $\left(\dfrac{3}{3} = 1, \text{ therefore } \dfrac{3a}{3} = 1a = a \right)$

(c) In this question, both sides of the equation will be multiplied by 5 to leave y on its own,

$\dfrac{y}{5} \times 5 = 3 \times 5$

$y = 15$ $\left(\dfrac{5}{5} = 1, \text{ therefore } \dfrac{y}{5} \times 5 = y \times 1 = y \right)$

(d) Transpose –7 to the RHS and –5x to the LHS to have the unknown terms on one side and the numbers on the other side.

$$2x - 3x + 5x = 3 + 7$$

$$7x - 3x = 10$$

$$4x = 10, \text{ or } x = \frac{10}{4} = \mathbf{2.5}$$

3.6 Application of linear equations

Equations can be used to solve many problems, simple as well as complex. The given information, which could be either in the written form or a sketch, can be used to form an equation

Example 3.7

A triangle has two unknown angles, as shown in Figure 3.1. Find the magnitudes of ∠A and ∠B.

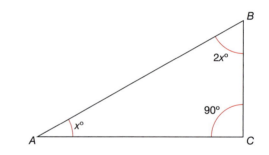

Figure 3.1

Solution:

The sum of the three angles of a triangle is 180°. Therefore:

$$x + 2x + 90° = 180°$$

$$3x + 90° = 180°$$

Transfer 90° to the RHS

$$3x = 180° - 90° = 90°$$

$$x = \frac{90}{3} = 30°$$

Therefore, **∠A = x = 30°** and **∠B = 2x = 60°**

Example 3.8

The length : width ratio of a rectangle is 2.5. If the perimeter of the rectangle is 42 cm, find its length and width.

Solution:

The perimeter of a rectangle is the sum of all the sides. The length : width ratio of 2.5 means that the length of the rectangle is 2.5 times the width. Assume the width of the rectangle to be x.

Length of the rectangle = $2.5x$

Perimeter = sum of the measurements of all sides

$$= x + x + 2.5x + 2.5x = 42$$

$7x = 42$

Therefore, $x = \dfrac{42}{7} = 6$ cm

Width of the rectangle = x = **6 cm**

Length = $2.5 \times$ width = 2.5×6 = **15 cm**

Exercise 3.1

The solutions to Exercise 3.1 can be found in Appendix 2.

1. Simplify (a) $5a + 2b + 3b - 2a$
   (b) $5x - 3y - 2x - 3y$.
2. Simplify $2xy^3 \times 5x^3y^3$.
3. Divide $25a^3b^2c^4$ by $5a^3bc^3$.
4. Simplify: (a) $4(2x + 3y)$
   (b) $2(3x - 6y)$
   (c) $5 + (x + 2y + 10)$
   (d) $3 - (-1x + 3y - 4z)$.
5. Solve the following equations:
   (a) $x - 5 = 14$
   (b) $2a = 15$
   (c) $\dfrac{y}{5} = 3.5$
   (d) $3(2x + 4) - 2(x - 3) = 4(2x + 4)$
   (e) $\dfrac{x}{2} + \dfrac{x+3}{4} = \dfrac{2x+1}{3}$.
6. A triangle has two unknown angles, $\angle A$ and $\angle B$. If $\angle A$ is twice the size of $\angle B$ and $\angle C = 75°$, find the magnitudes of $\angle A$ and $\angle B$.
7. The length : width ratio of a rectangle is 2.0. If the perimeter of the rectangle is 42 cm, find its length and width.
8. The length of a rectangle is 3 cm greater than its width. Obtain an expression for the perimeter of the rectangle by assuming the length to be x. Use the expression to calculate the length and width of a rectangle whose perimeter is 30 cm.
9. The angles of a quadrilateral ABCD are as shown in Figure 3.2. If the sum of the four angles is 360°, find the value of each angle.

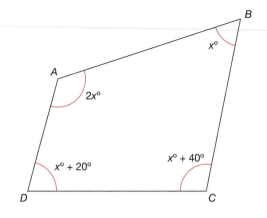

Figure 3.2

10. Three building plots are for sale in the Black Country. The price of plot A is £20 000 less than that of plot B, and the price of plot C is £30 000 more than that of plot B. Find the price of each plot if their total value is £190 000.

11. The outline plan of a building is shown in Figure 3.3. Find the lengths of the unknown sides if the perimeter of the building is 40 m.

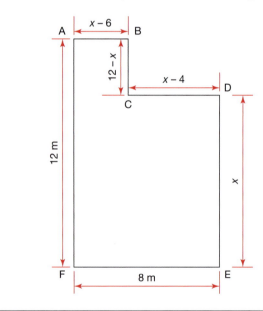

Figure 3.3

Answers to Exercise 3.1

1. (a) $3a + 5b$

   (b) $3x - 6y$

2. $10x^4 y^6$

3. $5bc$

4. (a) $8x + 12y$

(b) $6x - 12y$

(c) $x + 2y + 15$

(d) $3 + x - 3y + 4z$

5. (a) $x = 19$

(b) $a = 7.5$

(c) $y = 17.5$

(d) $x = 0.5$

(e) $x = -5$

6. $\angle A = 70°$, $\angle B = 35°$

7. Width = 7 cm; length = 14 cm

8. Width = 6 cm; length = 9 cm

9. $\angle A = 120°$, $\angle B = 60°$, $\angle C = 100°$, $\angle D = 80°$

10. Plot A = £40 000; Plot B = £60 000; Plot C = £90 000

11. AB = 3 m; BC = 3 m; CD = 5 m; DE = 9 m

CHAPTER 4

Indices and logarithms

Learning outcomes:

(a) Identify the laws of indices
(b) Perform calculations involving multiplication and division of powers
(c) Calculate logarithms and antilogarithms of numbers

4.1 Indices

This topic deals with numbers which are raised to a power. A question may involve a single number, or several numbers that are multiplied and/or divided. The rules that allow us to find solutions easily and quickly are called 'laws of indices', and are explained in Section 4.2.

Sometimes we come across a number which is multiplied by itself two times, three times or more. For example, 2 multiplied by itself four times can be written as $2 \times 2 \times 2 \times 2$. This can be written in a short form using indices:

$2 \times 2 \times 2 \times 2 = 2^4$ (two raised to the power 4)

2 is called the base, and 4 the index. The index tells us the number of times the base number is to be multiplied. Indices may be used when numbers and/or symbols are raised to a power, as shown in Example 4.1. The laws of indices are applicable to both.

4.2 Laws of indices

The laws of indices may be used to solve problems involving numbers raised to a power. They are applicable to negative and fractional indices in the same way as they are to positive whole-number indices.

4.2.1 Multiplication

When a number raised to a power is multiplied by the same number raised to a power, the indices are added:

$$a^m \times a^n = a^{m+n}$$

In this case, a is the base and m and n are the powers. It is important that in any one calculation the base numbers are the same.

Example 4.1

Simplify: (a) $3^2 \times 3^4$

(b) $a^3 \times a^2$

Solution:

(a) $a^m \times a^n = a^{m+n}$:

$$3^2 \times 3^4 = 3^{2+4} = 3^6 = \mathbf{729}$$

> **Proof:** 3^2, or 3 raised to the power 2, means that 3 is multiplied by itself twice. Similarly, 3^4 means that 3 is multiplied by itself 4 four times.
>
> $$3^2 \times 3^4 = (3 \times 3) \times (3 \times 3 \times 3 \times 3)$$
> $$= 3^6$$
> $$= \mathbf{729}$$

(b) $a^3 \times a^2 = a^{3+2} = \mathbf{a^5}$

Example 4.2

Simplify $5^2 \times 5^6 \times 5^4$

Solution:

The base numbers are the same in this question:

$$5^2 \times 5^6 \times 5^4 = 5^{2+6+4} = 5^{12} = \mathbf{244\ 140\ 625}$$

4.2.2 Division

In division the indices are subtracted if the base numbers are the same:

$$\frac{a^m}{a^n} = a^m \div a^n = a^{m-n}$$

Example 4.3

Simplify $\dfrac{7^6}{7^4}$

Solution:

The numerator and the denominator have the same base number, i.e. 7. Therefore we can use the law of indices:

$$\frac{7^6}{7^4} = 7^{6-4} = 7^2 = \textbf{49}$$

> **Proof:**
>
> $$\frac{7^6}{7^4} = \frac{7 \times 7 \times 7 \times 7 \times 7 \times 7}{7 \times 7 \times 7 \times 7} = 7 \times 7 = 7^2 = \textbf{49}$$

Example 4.4

Simplify $\dfrac{m^2 \times m^9}{m^3 \times m^6}$

Solution:

Use both laws (explained in Sections 4.2.1 and 4.2.2) to solve this question:

$$\frac{m^2 \times m^9}{m^3 \times m^6} = \frac{m^{2+9}}{m^{3+6}}$$

$$= \frac{m^{11}}{m^9}$$

$$= m^{11-9} = \textbf{\textit{m}}^{\textbf{2}}$$

4.2.3 Power of a power

If a number raised to a power is raised to another power, multiply the indices:

$$(a^m)^n = a^{mn}$$

Example 4.5

Simplify: (a) $(3^2)^4$

        (b) $(4a^2)^3$

Solution:

(a) $(3^2)^4$ means that 3^2 is multiplied by itself four times as the power (or index) is 4.

$(3^2)^4 = 3^2 \times 3^2 \times 3^2 \times 3^2 = 3^{2+2+2+2} = 3^8 = 6561$

Using the law $(a^m)^n = a^{mn}$, $(3^2)^4 = 3^{2 \times 4} = 3^8 = \textbf{6561}$

(b) $(4a^2)^3 = 4^3 \times a^{2 \times 3} = \textbf{64 } \textbf{\textit{a}}^{\textbf{6}}$

4.2.4 Negative powers

A simple example of a negative power is 2^{-3}, which can also be written as $\frac{1}{2^3}$. This can be written in a generalised form as:

$$a^{-n} = \frac{1}{a^n}$$

Example 4.6

Solve $\dfrac{2^3}{2^6}$

Solution:

Apply the division law $\dfrac{a^m}{a^n} = a^{m-n}$:

$$\frac{2^3}{2^6} = 2^{3-6} = 2^{-3} \qquad (1)$$

Also,

$$\frac{2^3}{2^6} = \frac{2 \times 2 \times 2}{2 \times 2 \times 2 \times 2 \times 2 \times 2}$$

$$= \frac{8}{8 \times 2 \times 2 \times 2} = \frac{1}{2 \times 2 \times 2} = \frac{1}{2^3} \qquad (2)$$

Answers (1) and (2) must be equal as they have resulted from the same problem, therefore, $\mathbf{2^{-3}} = \dfrac{1}{2^3}$

Example 4.7

Simplify $\dfrac{m^4}{m^3 \times m^3}$

Solution:

$$\frac{m^4}{m^3 \times m^3} = \frac{m^4}{m^{3+3}} = \frac{m^4}{m^6}$$

Apply the division rule, $\dfrac{m^4}{m^6} = m^{4-6}$

$$= \mathbf{m^{-2}} = \frac{1}{m^2}$$

4.2.5 Zero index

Any number raised to the power zero equals 1, or $a^0 = 1$. Example 4.8 verifies this law.

Example 4.8

Simplify $5^3 \div 5^3$

Solution:

$$5^3 \div 5^3 = \frac{5^3}{5^3} = 5^{3-3} = 5^0 \qquad (1)$$

Also,

$$\frac{5^3}{5^3} = \frac{5 \times 5 \times 5}{5 \times 5 \times 5} = \frac{125}{125} = 1 \qquad (2)$$

From (1) and (2), **$5^0 = 1$**

From Example 4.8 we can conclude that any number raised to the power zero is equal to 1.

$$(281)^0 = 1$$
$$(65\ 839)^0 = 1$$
$$x^0 = 1$$

4.3 Logarithms

Logarithms were essential tools to perform calculations involving multiplication, division, etc. before the use of electronic calculators. If a positive number x is expressed in the form: $x = a^y$, then y is called the logarithm of x to base a.

In mathematical terms: $y = \log_a x$, where log is the short form of logarithm. The most commonly used form of logarithms in mathematics and science are to the base 10. Use the $\boxed{\log}$ key on your calculator to determine the logarithm of a number.

Sometimes we know the logarithm of a number but have to find the number itself. This involves the reverse process of finding the logarithm of a number and is known as the antilogarithm. Use the shift/second function key before the $\boxed{\log}$ key to find the antilogarithm of a number.

Example 4.9

(a) Find the value of $\log_{10} 2.5 \times 10^5$
(b) Calculate the antilogarithm of 5.5

Solution:

Press the calculator keys in the following sequence:

(a) $\boxed{\log}$ $\boxed{2}$ $\boxed{.}$ $\boxed{5}$ $\boxed{\text{EXP}}$ $\boxed{5}$ $\boxed{=}$ **5.3979**

(b) $\boxed{\text{SHIFT}}$ $\boxed{\text{LOG}}$ $\boxed{5}$ $\boxed{.}$ $\boxed{5}$ $\boxed{=}$ **316 227.766**

Exercise 4.1

The solutions to Exercise 4.1 can be found in Appendix 2.

For questions 1–7, evaluate or simplify.

1. (a) $4^2 \times 4^7$ (b) $m^2 \times m^3$

2. (a) $3^2 \times 3^5 \times 3^7$

   (b) $4 \times 4^2 \times 4^7$

3. (a) $\dfrac{n^3 \times n^4}{n \times n^5}$

   (b) $\dfrac{x^6 \times x^2}{x^3 \times x}$

4. (a) $(2^3)^5$

   (b) $(3x^2)^4$

5. $\dfrac{3^4}{3^7}$

6. (a) $\dfrac{a^3}{a^2 \times a^5}$

   (b) $\dfrac{x^4 \times x^2}{x^3 \times x^5}$

7. (a) $\dfrac{a \times a^2 \times a^4}{a^3 \times a^4}$

   (b) $\dfrac{y^6 \times y^4}{y^3 \times y^5 \times y^2}$

8. Find the logarithm of (a) 25, (b) 150 and (c) 1204.

9. Find the logarithm of (a) 2.2×10^2 and (b) 3.8×10^{-3}.

10. Find the antilogarithm of (a) 8.5, (b) 0.72 and (c) 0.0014.

Answers to Exercise 4.1

1. (a) 4^9 or 262 144; (b) m^5
2. (a) 3^{14} or 4 782 969; (b) 4^{10} or 1 048 576
3. (a) n; (b) x^4
4. (a) 2^{15} or 32 768; (b) $81x^8$
5. $\dfrac{1}{3^3}$ or $\dfrac{1}{27}$
6. (a) $\dfrac{1}{a^4}$ or a^{-4}; (b) $\dfrac{1}{x^2}$ or x^{-2}
7. (a) 1; (b) 1
8. (a) 1.3979; (b) 2.1761; (c) 3.0806
9. (a) 2.3424; (b) −2.4202
10. (a) 316 227 766; (b) 5.248; (c) 1.003

Standard form, significant figures and estimation

Learning outcomes:

(a) Write a number in the standard form
(b) Write a number to any number of significant figures
(c) Estimate the answer of a calculation

5.1 Standard form

In scientific calculations, it is more convenient to write large or small numbers in a form known as the standard form. In the standard form a number is split into two parts:

(a) a decimal number that is greater than 1 but less than 10; and

(b) 10^n, where n could be a positive or negative integer, depending on the original number.

For example, 3530 is written in the standard form as:

$3530 = 3.530 \times 10^3$

As a starting point 3530 can be written as 3530.0

The first part (a) should be more than 1.0 but less than 10. Therefore move the decimal point by three places to the left to achieve this:

$3\ 5\ 3\ 0\ .\ 0 = 3.5300 \times b$ (3530.0 becomes 3.5300 after moving the decimal)

The second part (b) is a number that we must introduce to keep the value of 3530 unchanged. Since we have moved the decimal by three places to the left (or reduced the value of 3530), therefore, we must multiply 3.5300 by 1000. The three zeros in 1000 correspond to the number of places that the decimal has been moved.

$3.5300 \times b = 3.5300 \times 1000$

1000 can further be written as $10 \times 10 \times 10$ or 10^3. Therefore,

$$3.5300 \times 1000 = 3.5300 \times 10^3 = 3.53 \times 10^3 \quad (3.5300 = 3.53)$$

Another method of representing 3530 (which is not much different from the above) in the standard form is to divide and multiply it by 1000. This will not change the value of 3530 and at the same time $\dfrac{3530}{1000}$ will result in a number that is greater than 1 but less than 10.

$$3530 = \dfrac{3530}{1000} \times 1000$$
$$= 3.530 \times 1000 = 3.530 \times 10^3 \text{ or } 3.53 \times 10^3$$

A similar approach can be used if a number is less than 1. Consider 0.05, which can be written in a fractional form as:

$$0.05 = \dfrac{05}{100} = \dfrac{5}{100}$$
$$\dfrac{5}{100} = \dfrac{5}{10^2} = 5 \times 10^{-2} \quad \left(\dfrac{1}{10^2} = 10^{-2} \right)$$

Example 5.1

Write the following numbers in the standard form: (a) 425; (b) 15 230; (c) 0.0056; (d) 0.0000621

Solution:

(a) 425 can also be written as 425.0:

$$425 = 425.0 = 4.25 \, 0 \times 100 = 4.25 \times 10^2$$

(b) 15 230 can be written as 15230.0:

$$15230 = 15230.0 = 1.52300 \times 10000 = 1.52300 \times 10^4$$
$$= 1.523 \times 10^4$$

(c) $0.0056 = \dfrac{0005.6}{1000} = \dfrac{5.6}{10^3} = 5.6 \times 10^{-3}$

In this question the decimal point has been moved to the right by three places. Division by 1000 is necessary to keep the number unchanged. The three zeros in 1000 correspond to the number of places that the decimal has been moved.

(d) $0.0000621 = \dfrac{000006.21}{100000} = \dfrac{6.21}{10^5} = 6.21 \times 10^{-5}$

5.2 Significant figures

This is a method of approximating a number so that its value is not much different from the approximated value.

Consider a decimal number, say 8.354; it has been represented to four significant figures (s.f.). To reduce it to 3 s.f., the last figure has to be discarded. This process may affect the next figure, depending on the value of the discarded figure. If the last figure is between 0 and 4, the next figure remains the same. If, however, the last figure is 5 or greater, the next figure is increased by 1.

The last figure in 8.354 is 4. As it is less than 5, the next figure remains unchanged.

8.354 = 8.35   correct to 3 s.f.

If we want to reduce 8.35 to two significant figures, again the last figure will be discarded. As the value of the last figure is 5, the next figure will be increased by 1.

8.35 = 8.4   correct to 2 s.f.

Similarly,

8.4 = 8   correct to 1 s.f.

In very small decimal numbers, e.g. 0.005739, the zeros after the decimal point are significant figures and must be kept during the approximation process. Example 5.2 illustrates this point.

Example 5.2

Write: (a) 12.268 5, correct to five, four, three and two significant figures.

(b) 0.005739 correct to three, two and one significant figures.

Solution:

(a) 12.2685 has six significant figures and is slightly more than 12. Discarding figures after the decimal point will not have a major effect on the value of 12.2685. Each time the last figure is discarded, the number of significant figures is reduced by one.

12.2685 = **12.269**   correct to 5 s.f.

= **12.27**   correct to 4 s.f.

= **12.3**   correct to 3 s.f.

= **12**   correct to 2 s.f.

(b)

0.005739 = **0.00574** correct to 3 s.f.

= **0.0057**   correct to 2 s.f.

= **0.006**   correct to 1 s.f.

Example 5.3

Write 478 353 to five, four, three, two and one significant figures.

Solution:

The solution of this example is different from the previous one as 478 353 is an integer (whole number). The last figure cannot be discarded as this will change the value of the number completely. Instead, the procedure involves replacing the last figure by zero:

$$478\ 353 = \mathbf{478\ 350} \quad \text{correct to 5 s.f.}$$
$$= \mathbf{478\ 400} \quad \text{correct to 4 s.f.}$$
$$= \mathbf{478\ 000} \quad \text{correct to 3 s.f.}$$
$$= \mathbf{480\ 000} \quad \text{correct to 2 s.f.}$$
$$= \mathbf{500\ 000} \quad \text{correct to 1 s.f.}$$

5.3 Estimation

Estimation is the process of finding an approximate answer to a question. It can be used to check if the actual answer is right, as sometimes the data input on a calculator may not be right if a wrong key is pressed. Estimation is done by rounding the figures, which makes addition, multiplication, division, etc. simpler and quicker. For example, 210 can be rounded to 200 (nearest 100)

Example 5.4

(a) Estimate the results of: (1) 263 + 187 + 221

  (2) 23 × 13

  (3) $\dfrac{27 \times 53}{12 \times 18}$

(b) Solve the above questions using a calculator and compare the results.

Solution:

(a) (1) To estimate the answer to 263 + 187 + 221, round the figures and add:

250 + 200 + 200 = **650**

(2) After rounding the figures, 23 × 13 becomes:

25 × 10 = **250**

(3) $\dfrac{27 \times 53}{12 \times 18}$ can be rounded to:

$\dfrac{30 \times 50}{10 \times 20} = \dfrac{15}{2} = \mathbf{7.5}$

(b) (1) 263 + 187 + 221 = **671**

(2) 23 × 13 = **299**

(3) $\dfrac{27 \times 53}{12 \times 18} = \mathbf{6.625}$

A comparison of the estimated and the actual answers is given in Table 5.1

Table 5.1

Question	Estimated answer	Actual answer
263 + 187 + 221	650	671
23 × 13	250	299
$\dfrac{27 \times 53}{12 \times 18}$	7.5	6.625

Exercise 5.1

The solutions to Exercise 5.1 can be found in Appendix 2.
   For questions 1–2 write the numbers in standard form.

1. (a) 976; (b) 1478; (c) 377 620

2. (a) 0.025; (b) 0.00071; (c) 0.000000437

3. Write the following standard forms as ordinary numbers: (a) 1.721×10^2; (b) 2.371×10^{-3}; (c) 9.877×10^4; (d) 9.1×10^{-6}.

4. Write 361.7297 correct to five, four, three and two significant figures.

5. Write 867 364 correct to five, four, three, two and one significant figures.

6. Write 0.000839 correct to two and one significant figures.

7. Estimate the results of the following and draw a table comparing your estimated answers with the accurate answers: (a) 462 + 122 + 768; (b) 38 × 15; (c) $\dfrac{61 \times 89}{11 \times 29}$.

Answers to Exercise 5.1

1. (a) 9.76×10^2; (b) 1.478×10^3; (c) 3.7762×10^5

2. (a) 2.5×10^{-2}; (b) 7.1×10^{-4}; (c) 4.37×10^{-7}

3. (a) 172.1; (b) 0.002371; (c) 98 770; (d) 0.0000091

4. 361.73, 361.7, 362, 360

5. 867 360, 867 400, 867 000, 870 000, 900 000

6. 0.00084, 0.0008

7.

Question	Estimated answer	Actual answer
462 + 122 + 768	1350	1352
38 × 15	600	570
$\dfrac{61 \times 89}{11 \times 29}$	18	17.019

CHAPTER 6

Transposition and evaluation of formulae

Learning outcomes:

(a) Transpose simple formulae involving one or more of addition, subtraction, multiplication, division, squares or square roots

(b) Evaluate simple formulae

6.1 Transposition of formulae

Formulae are used in mathematics and science to work out solutions to many problems. For example, the area of a circle is given by:

Area, $A = \pi r^2$, where π is a constant and r is the radius of the circle.

In this formula, A is called the subject of the formula, and can be evaluated by putting in the values of π and the radius.

There might be a situation where the area of a circle is given and we are asked to calculate the radius. Before the calculations are performed it is better to rearrange the formula to make r the subject. This process of rearranging formulae is called the transposition of formulae.

The method of transposition depends on the given formula. Three types of formula are considered here:

Type 1 the components of a formula are added and/or subtracted

Type 2 the components of a formula are multiplied and/or divided

Type 3 a combination of Type 1 and Type 2 formulae.

6.1.1 Type 1 formulae

Consider the formula $a = b - c + d$

A formula is an equation. The 'equal to' sign (=) creates the left-hand side (LHS) and the right-hand side (RHS). If we want to make b the subject of the formula, the method will involve rearrangement so that it is:

- on the LHS of the formula
- on its own
- on the numerator side.

When a number or a symbol is moved to the other side of the equation, its sign will change from + to −, and from − to +. The objective in this example is to rearrange c and d so that in the end there is only b on one side of the equation.

Take c to the LHS; its sign will change from − to +:

$$a + c = b + d$$

Take d to the LHS; its sign will change from + to −:

$$a + c - d = b$$

As the LHS is equal to the RHS, this equation can be written as:

$$b = a + c - d$$

6.1.2 Type 2 formulae

Consider the formula for determining the volume of a cuboid:

Volume, $V = L \times W \times H$

where L, W and H are the length, width and height, respectively. If we want to make L the subject of the formula, then we need to transfer W and H to the LHS. This can be achieved by dividing both sides by $W \times H$.

$$\frac{V}{W \times H} = \frac{L \times W \times H}{W \times H}$$

$$\frac{V}{W \times H} = L \times 1 \times 1 \quad \left(\frac{W}{W} = 1; \; \frac{H}{H} = 1 \right)$$

$$\text{or } L = \frac{V}{W \times H}$$

6.1.3 Type 3 formulae

The formulae in this category are combinations of Type 1 and Type 2. For example:

$$y = mx + c$$

Suppose we are required to make m the subject of the formula. To achieve this, first transfer c to the LHS and then transpose x as explained in Sections 6.1.1 and 6.1.2. On transferring c to the LHS,

$$y - c = mx$$

As m and x are multiplied, the only way to have just m on the RHS is to divide both sides by x:

$$\frac{y-c}{x} = \frac{mx}{x}$$

$$\frac{y-c}{x} = m$$

or $m = \dfrac{y-c}{x}$

Example 6.1

Transpose $a + 2c - b = d + c - a$ to make d the subject.

Solution:

Transfer $-a$ and $+c$ to the LHS. Their signs will change:

$a + a + 2c - c - b = d$

$2a + c - b = d$

or $\boldsymbol{d = 2a + c - b}$

Example 6.2

Transpose $c = 2\pi r$ to make r the subject

Solution:

Divide both sides by 2π. This will leave only r on the RHS and hence it becomes the subject.

$$\frac{c}{2\pi} = \frac{2\pi r}{2\pi}$$

$$\frac{c}{2\pi} = r$$

$$\therefore r = \frac{c}{2\pi}$$

Example 6.3

Transpose: (a) $A = \pi r^2$ to make r the subject

(b) $r = \sqrt{\dfrac{V}{\pi h}}$ to make V the subject

Solution:

(a) $A = \pi r^2$ can be written as $\pi r^2 = A$. Divide both sides by π to have only r^2 on the LHS:

$$\frac{\pi r^2}{\pi} = \frac{A}{\pi}$$

$$r^2 = \frac{A}{\pi}$$

Take square root of both sides. As the square root is the reverse of squaring, the square and the square root cancel out to yield r:

$$\sqrt{r^2} = \sqrt{\frac{A}{\pi}}$$

$$\therefore r = \sqrt{\frac{A}{\pi}}$$

(b)

$$r = \sqrt{\frac{V}{\pi h}} \quad \text{can be written as} \quad \sqrt{\frac{V}{\pi h}} = r$$

Square both sides to get rid of the square root, as explained before

$$\frac{V}{\pi h} = r^2$$

Multiply both sides by πh:

$$\frac{V \pi h}{\pi h} = r^2 \pi h, \text{ or } \mathbf{V = r^2 \pi h}$$

Example 6.4

Transpose $F = \dfrac{9}{5}C + 32$ to make C the subject

Solution:

Transpose +32 to the LHS. Its sign will change from + to −:

$$F - 32 = \frac{9}{5}C$$

Multiply both sides by 5:

$$5(F - 32) = 5 \times \frac{9}{5}C$$

$$5(F - 32) = 9C$$

Divide both sides by 9 to get only C on the RHS

$$\frac{5}{9}(F - 32) = \frac{9}{9}C$$

$$\frac{5}{9}(F - 32) = C$$

$$\text{or } \mathbf{C = \frac{5}{9}(F - 32)}$$

6.2 Evaluation of formulae

Evaluation of formulae involves quantifying one term in a formula by replacing the other terms by their given values. For example, the area of a rectangle:

Area = length × width

If the length and the width of a rectangle are 10 cm and 6 cm, respectively, the area of the rectangle is:

$A = 10 \times 6 = 60$ cm^2

It is essential that the units of different quantities are compatible, otherwise the answer will be wrong. For instance, the length of a rectangle could be in metres and the width in centimetres. In this case it is necessary to convert either the length into centimetres or the width into metres. Example 6.5 illustrates this point.

Example 6.5

The surface area, S, of a circular column, excluding the ends, is given by $S = 2\pi rh$, where r and h are the radius and the height, respectively. Calculate the surface area of the column in cm^2 if $r = 30$ cm and $h = 3.0$ m.

Solution:

As the units of r and h are not similar, and because the answer is required in cm^2, convert 3.0 m into centimetres.

$h = 3.0$ m $\times 100 = 300$ cm $\qquad$ (1 m = 100 cm)

$S = 2\pi rh = 2 \times \pi \times 30 \times 300$

$\qquad$ = **56 548.67 cm^2**

Example 6.6

If $C = -10°$, calculate F in the formula $F = \dfrac{9}{5}C + 32$.

Solution:

$F = \dfrac{9}{5}(-10) + 32$

$\qquad = -18 + 32 =$ **14°**

Example 6.7

Evaluate r in the formula $r = \sqrt{\dfrac{s}{4\pi}}$, if $s = 400$ cm^2

Solution:

$r = \sqrt{\dfrac{s}{4\pi}}$

$r = \sqrt{\dfrac{400}{4\pi}}$

$r = \sqrt{31.827} =$ **5.64 cm**

6.3 Evaluation of formulae: practical examples

Example 6.7

Figure 6.1 shows two resistors connected in parallel in a circuit. Calculate the total resistance (R) of the circuit, if $\dfrac{1}{R} = \dfrac{1}{R_1} + \dfrac{1}{R_2}$

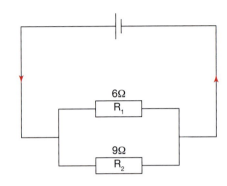

Figure 6.1

Solution:

$$\frac{1}{R} = \frac{1}{R_1} + \frac{1}{R_2}$$

From Figure 6.1, $R_1 = 6\ \Omega$ and $R_2 = 9\ \Omega$:

$$\frac{1}{R} = \frac{1}{6} + \frac{1}{9} = \frac{3+2}{18} = \frac{5}{18}$$

After transposition, $5R = 18$, or $R = \dfrac{18}{5} = \mathbf{3.6}\ \Omega$ (Ohm (Ω) is the unit of electrical resistance).

Example 6.8

A transformer steps down the mains supply of 230 V to 110 V to operate power tools. If the secondary coil of the transformer has 88 turns, use the following formula to calculate the number of turns on the primary coil:

$$\frac{V_p}{V_s} = \frac{N_p}{N_s}$$

Solution:

Transformers are used to increase/decrease the voltage so that some electrical devices can work using the mains supply. For example, the mains voltage of 230 V is reduced for using calculators, laptops, power tools, etc.

The following formula can be used to solve this question:

$$\frac{\text{Primary voltage } (V_p)}{\text{Secondary voltage } (V_s)} = \frac{\text{Number of turns in the primary coil } (N_p)}{\text{Number of turns in the secondary coil } (N_s)}$$

$$\frac{230}{110} = \frac{N_p}{88}$$

or $N_p = \dfrac{230 \times 88}{110} = \textbf{184}$

Example 6.9

The volume of internal air in a building is 600 m³, which is to be maintained at a temperature of 22°C while the external air temperature is 2°C. Calculate the rate of heat loss due to ventilation if one air change is to be allowed per hour and the specific heat capacity of air is 1200 J/m³ °C.

Solution:

The rate of heat loss due to ventilation is:

$$\frac{C_v \times V \times N \times T}{3600}$$

where C_v is the volumetric specific heat capacity of air, V is the volume of the internal air, N is the number of air changes per hour and T is the difference between the internal and external temperatures.

C_v = 1200 J/m³ °C; V = 600 m³; N = 1 per hour; T = 22 – 2 = 20°C.

Rate of heat loss due to ventilation = $\dfrac{1200 \times 600 \times 1 \times 20}{3600}$ = **4000 W**

Example 6.10

A cold water storage cistern feeds water to a vessel as shown in Figure 6.2.

(a) Use the formula, $p = \rho g h$ to calculate the pressure (p) on the base of the vessel.

(b) Calculate the force acting on the base of the vessel, if force = pressure × area.

The density of water (ρ) is 1000 kg/m³, and the acceleration due to gravity (g) is 9.81 ms⁻².

Solution:

(a) Water pressure at the base of the vessel,

$p = \rho g h$

$= 1000 \times 9.81 \times 3.2$

$= \textbf{31 392 N/m}^2 \textbf{ or 31 392 Pa}$

(Note: 1 Newton/m² = 1 Pascal (Pa))

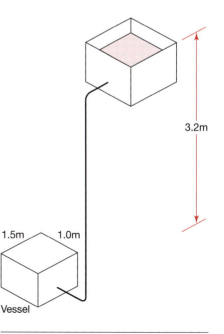

1.5m   1.0m

Vessel

3.2m

Figure 6.2

(b)

$$\text{force} = \text{pressure} \times \text{area}$$
$$= 31\ 392 \times (1.5 \times 1)$$
$$= \textbf{47 088 N or 47.088 kN}$$

Example 6.11

A circular drain of 300 mm diameter runs half-full of water. Use Chézy's formula to find the velocity of flow if the drain is laid at a gradient of 1 in 100. The value of Chézy coefficient (c) is 50 m$^{1/2}$/s.

Solution:

Chézy's formula is used to find the velocity (v) of liquid flow in drains and partly filled pipes:

$v = c \sqrt{R\,S}$, where c is the Chézy coefficient, R is the hydraulic radius and S is the slope (or gradient).

In this case, the radius of the drain is 150 mm or 0.150 m:

$$\text{Cross-sectional area of flow} = \frac{\pi r^2}{2} \text{ (the drain runs half-full)}$$

$$= \frac{\pi \times (0.150)^2}{2} = 0.03534 \text{ m}^2$$

$$\text{Wetted perimeter} = \frac{2\pi r}{2} \text{ (half perimeter as the drain runs half-full)}$$

$$= \frac{2 \times \pi \times 0.150}{2} = 0.47124 \text{ m}$$

$$\text{Hydraulic radius, } R = \frac{\text{Cross-sectional area of flow}}{\text{Wetted perimeter}}$$

$$= \frac{0.03534}{0.47124} = 0.075$$

Slope, S = 1 in 100

$$= \frac{1}{100} = 0.01$$

$$\text{Velocity of flow, } v = c\sqrt{R\,S} = 50\sqrt{0.075 \times 0.01}$$
$$= 50 \times 0.02739 = \textbf{1.37 m/s}$$

Example 6.12

A street lamp, having a luminous intensity (I) of 750 candela (cd), is suspended 5.5 m above the edge of a road, as shown in Figure 6.3. If the road is 6.0 m wide, calculate the illuminance (E):

(a) At point B, which is directly under the lamp;

(b) At point C, which is directly opposite B, but on the other edge of the road. Illuminance directly under the lamp: $E = \dfrac{I}{h^2}$ and illuminance at

other points: $E = \dfrac{I}{h^2} \cos^3\theta$

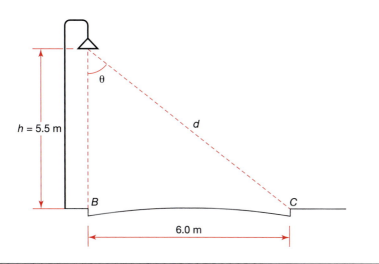

Figure 6.3

Solution:

(a) At point B:

$$\text{Illuminance, } E = \frac{I}{h^2} = \frac{750}{5.5^2} = \textbf{24.8 lux}$$

(b) From Figure 6.3, $d^2 = 5.5^2 + 6^2$. Therefore,

$$d = \sqrt{66.25} = 8.1394 \text{ m}$$

$$\cos\theta = \frac{5.5}{8.1394} = 0.6757$$

$$E = \frac{I}{h^2}\cos^3\theta = \frac{750}{5.5^2} \times (0.6757)^3$$

$$= \textbf{7.7 lux}$$

Exercise 6.1

The solutions to Exercise 6.1 can be found in Appendix 2.

1. Transpose $b = c - d - a$, to make a the subject.
2. (a) Transpose $y = mx + c$, to make c the subject.
   (b) Calculate c if $y = 10$, $x = 2$ and $m = 4$.
3. Transpose $v = u + at$, to make t the subject.
4. Transpose $p = \dfrac{\pi d}{2}$ to make d the subject.
5. (a) The flow of heat (Q) through a material is given by, $Q = \dfrac{kA\theta}{d}$.
   Transpose the formula to make k the subject.
   (b) Find the value of k if $Q = 2000$, $A = 10$, $\theta = 40$ and $d = 0.5$.
6. (a) The surface area (A) of an object is given by, $A = 4\pi r^2$. Transpose the formula to make r the subject.
   (b) Find the value of r if $A = 5000$ cm^2.

7. The volume of a cylinder is given by $V = \pi r^2 h$. Transpose the formula:

   (a) to make h the subject

   (b) to make r the subject.

8. (a) Transpose $v^2 - u^2 = 2as$, to make s the subject of the formula.

   (b) Find the value of s if $v = 5$ m/s, $u = 0$ and $a = 2.5$ m/s^2.

9. Transpose $y = mx + c$, to make x the subject.

10. Transpose $y = \dfrac{3x + 2z}{7} + d$, to make x the subject.

11. Transpose $V = \dfrac{\pi}{3} r^2 h$, to make r the subject.

12. The velocity (v) of water flowing in an open channel is given by $v = c\sqrt{mi}$.

   (a) Transpose the formula to make m the subject.

   (b) Calculate m if $v = 2.42$ m/s, $c = 50$ and $i = 0.025$.

13. A 5 Ω resistor and 10 Ω resistor are connected in parallel in a circuit. Calculate the total resistance (R) of the circuit. Refer to Example 6.7 for the formula.

14. The rate of heat loss due to ventilation is given by the following formula:

   $$\text{Rate of heat loss due to ventilation} = \frac{C_v \times V \times N \times T}{3600}$$

   If $C_v = 1212$ J/m^3 °C, $V = 400$ m^3, $N = 2.0$ and $T = 21$°C, calculate the rate of heat loss due to ventilation.

15. Calculate the heat energy and power required to raise the temperature of 162 litres of water from 10°C to 50°C:

   Heat energy $= MST$, where M is the mass of water, S is the specific heat capacity of water and T is the temperature difference

   $$\text{Power: 1 kW/hr} = \frac{1 \text{ kilojoule}}{1 \text{ second}} \times 3600 \text{ seconds}$$

   The specific heat capacity of water is 4.186 kJ/kg °C.

16. A cold water storage cistern (CWSC) feeds water to a vessel, which is sited 4.0 m below the CWSC. Calculate the force exerted by water on the base of the vessel that measures 1.5 m × 1.2 m. Use the formulae given in Example 6.10.

17. (a) A circular drain of 150 mm diameter runs half-full of water. Use Chézy's formula ($v = c\sqrt{RS}$) to find the velocity of flow if the drain is laid at a gradient of 1 in 80. The values of Chézy coefficient (c) and hydraulic radius (R) are 50 m$^{1/2}$/s and 0.0375, respectively.

   (b) If A represents the cross-sectional area (m^2) of flow and v represents the flow velocity (m/s), calculate the flow rate (Q) in m^3/s from the following formula:

   $$Q = A \times v$$

Answers to Exercise 6.1

1. $a = c - d - b$

2. (a) $c = y - mx$; (b) 2

3. $t = \dfrac{v - u}{a}$

4. $d = \dfrac{2p}{\pi}$

5. (a) $k = \dfrac{Q\,d}{A\,\theta}$; (b) 2.5

6. (a) $r = \sqrt{\dfrac{A}{4\pi}}$; (b) 19.95 cm

7. (a) $h = \dfrac{V}{\pi r^2}$; (b) $r = \sqrt{\dfrac{V}{\pi h}}$

8. (a) $s = \dfrac{v^2 - u^2}{2a}$; (b) 5 m

9. $x = \dfrac{y - c}{m}$

10. $x = \dfrac{7y - 2z - 7d}{3}$

11. $r = \sqrt{\dfrac{3V}{\pi h}}$

12. (a) $m = \dfrac{v^2}{c^2 i}$; (b) 0.0937

13. 3.33 Ω

14. 5656 W

15. Heat energy required = 27 125.3 kJ; power required = 7.53 kW

16. (a) 39 240 Pa; (b) 70 632 N or 70.632 kN

17. (a) 1.53 m/s; (b) 0.0135 m³/s

CHAPTER **7**

Fractions and percentages

Learning outcomes:

(a) Add, subtract, multiply and divide fractions
(b) Convert fractions into decimals and percentages, and vice versa
(c) Calculate percentages and bulking of soils

7.1 Fractions

Consider a circle divided into five equal parts, as shown in Figure 7.1

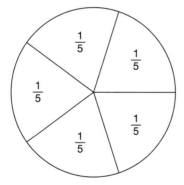

Figure 7.1

Each part is $\frac{1}{5}$ of the circle. This method of expressing the result, i.e. $\frac{1}{5}$, is called a fraction. Similarly:

two parts $= \frac{1}{5} \times 2 = \frac{2}{5}$ of the circle (Figure 7.2a)

three parts $= \frac{1}{5} \times 3 = \frac{3}{5}$ of the circle (Figure 7.2b)

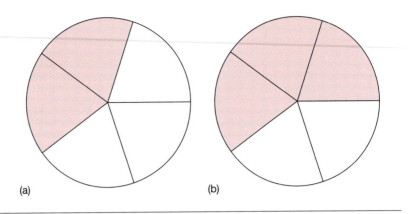

(a) (b)

Figure 7.2

The top number in a fraction is called the numerator; the bottom number is the denominator. If we consider the fraction $\frac{2}{5}$, 2 is the numerator and 5 the denominator.

Fractions can also be written in words, for example:

$\frac{1}{2}$ – half $\frac{1}{3}$ – one-third

$\frac{2}{3}$ – two-thirds $\frac{1}{4}$ – one-quarter

$\frac{3}{4}$ – three-quarters

Example 7.1

Express the shaded portions of the shapes shown in Figure 7.3 as a fraction.

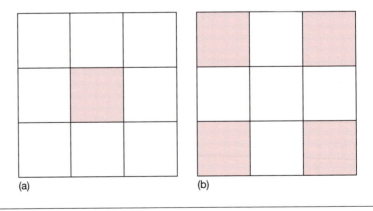

(a) (b)

Figure 7.3

Solution:

(a) Total number of squares = 9; shaded square = 1; each square is $\frac{1}{9}$ of the shape.

Therefore, the shaded portion = $\frac{1}{9}$ of the shape.

The above method can also be expressed as:

$$\text{Answer (as a fraction)} = \frac{\text{Shaded part/parts}}{\text{Total number of parts}} = \frac{1}{9}$$

(b) In this case: the number of shaded squares = 4; total number of squares = 9

$$\text{Answer (as a fraction)} = \frac{\text{Shaded squares}}{\text{Total number of squares}} = \frac{4}{9}$$

Example 7.2

Calculate (a) $\frac{2}{7}$ of 35; (b) $\frac{2}{3}$ of £19.80

Solution:

To solve these questions, replace 'of' with multiplication and simplify.

(a)

$$\frac{2}{7} \text{ of } 35 = \frac{2}{7} \times 35$$

$$= \frac{2 \times 35}{7} = \frac{70}{7} = \mathbf{10}$$

(b)

$$\frac{2}{3} \text{ of } 19.80 = \frac{2}{3} \times 19.80$$

$$= \frac{2 \times 19.80}{3} = \frac{39.60}{3} = \mathbf{13.20}$$

7.1.1 Simplification of fractions

In some fractions the numerator and the denominator can both be divided by the same number (known as the common factor), resulting in a simplified answer.

For example, in the fraction $\frac{2}{6}$, both 2 and 6 can be divided by 2:

$$\frac{2}{6} = \frac{2 \div 2}{6 \div 2} = \frac{1}{3}$$

$\frac{1}{3}$ cannot be further simplified, hence the fraction is said to be in its lowest terms.

Example 7.3

Express 20 pence as a fraction of 90 pence.

Solution:

As explained in Example 7.1, 20 pence as a fraction of 90 pence is $\frac{20}{90}$. $\frac{20}{90}$ can be simplified by dividing 20 and 90 by 10, which is a common factor:

$$\frac{20}{90} = \frac{20 \div 10}{90 \div 10} = \frac{2}{9}$$

Example 7.4

A patio consists of four red and eight grey slabs. Express the number of red slabs as a fraction of the total slabs.

Solution:

Total number of slabs = 4 + 8 = 12

$$\text{Red slabs as a fraction of 12 slabs} = \frac{\text{Number of red slabs}}{\text{Total number of slabs}}$$

$$= \frac{4}{12}$$

To reduce $\frac{4}{12}$ to its lowest terms, divide 4 and 12 by their common factor, i.e. 4:

$$\frac{4}{12} = \frac{4 \div 4}{12 \div 4} = \frac{1}{3}$$

Example 7.5

The components of a mortar mix and their masses are shown in Figure 7.4. Express the mass of lime as a fraction of the total mass of the mortar.

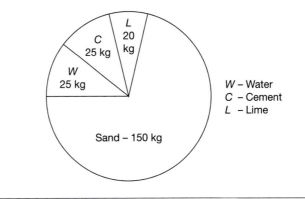

Figure 7.4

Solution:

$$\text{Mass of lime as a fraction of the total mass} = \frac{\text{Mass of lime}}{\text{Total mass of mortar}}$$

$$= \frac{20}{25 + 20 + 150 + 25}$$

$$= \frac{20}{220}$$

Divide 20 and 220 by 10, their common factor:

$$\frac{20}{220} = \frac{20 \div 10}{220 \div 10} = \frac{2}{22}$$

$\dfrac{2}{22}$ can be further simplified:

$$\frac{2}{22} = \frac{2 \div 2}{22 \div 2} = \frac{1}{11}$$

7.1.2 Equivalent fractions

Two equal fractions are known as equivalent fractions. In Example 7.4, $\dfrac{4}{12}$ results in $\dfrac{1}{3}$ after simplification, which means that $\dfrac{4}{12}$ is equal to $\dfrac{1}{3}$ as the simplification process is used only to express the fraction in its lowest terms. The equality of $\dfrac{4}{12}$ and $\dfrac{1}{3}$ can be verified by converting these fractions to decimals. Some examples of equivalent fractions are:

$\dfrac{2}{3}$ is equivalent to $\dfrac{10}{15}$ $\quad \left(\dfrac{10}{15} = \dfrac{10 \div 5}{15 \div 5} = \dfrac{2}{3} \right)$

$\dfrac{1}{2}$ is equivalent to $\dfrac{3}{6}$ $\quad \left(\dfrac{3}{6} = \dfrac{3 \div 3}{6 \div 3} = \dfrac{1}{2} \right)$

7.1.3 Addition and subtraction of fractions

Fractions can be added/subtracted using a range of methods. One of these methods is explained here:

Step 1: Multiply and divide each fraction by a suitable number to make the denominators of the fractions equal. Although different numbers can be used for different fractions, multiply and divide a fraction by the same number.

Step 2: Add or subtract the numerators and divide by the denominator.

Step 3: Simplify the fraction, if necessary, to bring it to its lowest terms. Example 7.6 illustrates this procedure.

Example 7.6

Add $\dfrac{1}{3}$ and $\dfrac{2}{5}$

Solution:

Step 1: Multiply and divide $\dfrac{1}{3}$ by 5 and $\dfrac{2}{5}$ by 3 to make the denominator in both fractions equal:

$$\frac{1}{3} \times \frac{5}{5} = \frac{5}{15}$$

$$\frac{2}{5} \times \frac{3}{3} = \frac{6}{15}$$

Step 2:

$$\frac{5}{15} + \frac{6}{15} = \frac{5+6}{15} = \frac{11}{15}$$

Step 3: $\dfrac{11}{15}$ cannot be simplified as there are no common factors.

Therefore, $\dfrac{11}{15}$ is the answer.

Example 7.7

Subtract $\dfrac{2}{3}$ from $\dfrac{3}{4}$

Solution:

Using mathematical symbols, the question is written as: $\dfrac{3}{4} - \dfrac{2}{3}$

Step 1: Multiply and divide $\dfrac{3}{4}$ by 3 and $\dfrac{2}{3}$ by 4 to make the denominators equal.

$$\frac{3}{4} \times \frac{3}{3} = \frac{9}{12}$$

$$\frac{2}{3} \times \frac{4}{4} = \frac{8}{12}$$

Step 2:

$$\frac{9}{12} - \frac{8}{12} = \frac{9-8}{12} = \frac{1}{12}$$

Step 3: $\dfrac{1}{12}$ cannot be simplified further, and hence is the answer.

7.1.4 Multiplication and division of fractions

In order to multiply two or more fractions:

Step 1: Multiply the top numbers

Step 2: Multiply the bottom numbers

Step 3: Simplify the answer to its lowest terms.

In division, the dividing fraction is inverted to change division into multiplication. For example, 5 divided by 6 = $5 \div 6 = 5 \times \dfrac{1}{6}$. Here 6 is inverted as division is changed into multiplication. Similarly, $\dfrac{1}{3} \div \dfrac{2}{3} = \dfrac{1}{3} \times \dfrac{3}{2}$.

Example 7.8

(a) Multiply $\dfrac{3}{7}$ and $\dfrac{1}{2}$; (b) divide $\dfrac{3}{7}$ by $\dfrac{1}{2}$

Solution:

(a) Steps 1 and 2:

$$\frac{3}{7} \times \frac{1}{2} = \frac{3 \times 1}{7 \times 2} = \frac{3}{14}$$

Step 3: $\dfrac{3}{14}$ is the answer as it cannot be simplified to lower terms

(b) The question can be written as $\dfrac{3}{7} \div \dfrac{1}{2}$. Invert the dividing fraction:

$$\frac{3}{7} \div \frac{1}{2} = \frac{3}{7} \times \frac{2}{1}$$

$$= \frac{3 \times 2}{7 \times 1} = \frac{6}{7}$$

7.1.5 Conversion of fractions to decimals

A fraction can be converted to a decimal number by dividing the numerator by the denominator. For example:

$$\frac{1}{5} = 1 \div 5 = 0.2 \quad \text{(for details on decimals see Chapter 2)}$$

Example 7.9

Convert into decimal numbers: (a) $\dfrac{1}{4}$; (b) $\dfrac{2}{5}$.

Solution:

(a) $\dfrac{1}{4} = 1 \div 4 = \mathbf{0.25}$;   (b) $\dfrac{2}{5} = 2 \div 5 = \mathbf{0.40}$

7.2 Percentages

Percentage calculation is a convenient and easily understandable method of comparing fractions, and forms an important topic of mathematics. The words 'per cent' have been taken from Latin, and mean 'per hundred'. Percentages are used in everyday life – for example, bank/building society interest rates, VAT and discounts in shops/stores are shown as percentages.

A discount of 10% (per cent is denoted as %) means that if a customer wants to buy goods worth £100, they will get a discount of £10, and hence pay £90.

7.2.1 Conversion of fractions and decimals into percentages

As per cent means 'per hundred' the conversion of either a fraction or a decimal can be achieved by multiplying them by 100 and vice versa. Examples 7.10–7.13 illustrate the procedure.

Example 7.10

Convert $\dfrac{3}{25}$ into a percentage.

Solution:

To convert a fraction into a percentage, multiply it by 100:

$$\frac{3}{25} = \frac{3}{25} \times 100 = \frac{300}{25} = \mathbf{12\%}$$

Example 7.11

Convert 0.25 into a percentage.

Solution:

Multiply the decimal number by 100 to convert it into a percentage:

$$0.25 = 0.25 \times 100 = \mathbf{25\%}$$

Example 7.12

A student has obtained 60 marks out of 75 in the science examination. What is her percentage mark?

Solution:

To solve this question, express 60 marks as a fraction of 75 and multiply by 100

$$60 \text{ marks out of } 75 = \frac{60}{75}$$

$$= \frac{60}{75} \times 100 = \mathbf{80\%}$$

Example 7.13

Convert 40% into: (a) a fraction; (b) a decimal

Solution:

40% means '40 per 100' or $\frac{40}{100}$.

(a) $\frac{40}{100}$ is a fraction which can be simplified further:

$$\frac{40}{100} = \frac{40 \div 20}{100 \div 20} = \frac{2}{5}$$

(b) $\frac{40}{100}$ can be written as $40 \div 100$

$$40 \div 100 = 0.4$$

7.2.2 Value added tax (VAT)

VAT is a sale tax which is added to the cost of most of the goods sold in shops, supermarkets, etc. The current rate of VAT is 20%, which means an item costing £100 without VAT will cost £120.00 with VAT. For other amounts the VAT will increase or decrease accordingly.

Example 7.14

The price of a chainsaw, displayed in a warehouse, is £109.95 excluding VAT. If VAT is charged at 20%, find the price of the chainsaw, inclusive of VAT.

Solution:

The amount of VAT = 20% of 109.95

$$= \frac{20}{100} \times 109.95 = £21.99$$

The price of the chainsaw, including VAT = £109.95 + £21.99 = **£131.94**.

7.3 Bulking of sand

Sand is one of the three types of soils which, as a subsoil, supports the weight of buildings and other structures. Sand is also used in the manufacture of concrete and cement/sand mortar. When concrete mixes are specified by volume, the sand is assumed to be dry. However, in many cases sand may not be completely dry. When dry sand comes into contact with a small amount of water, a thin film of moisture is formed around the particles. This causes the particles to move away from each other and results in an increase in the volume of sand. This phenomenon is known as the bulking of sand. Unless allowance is made for bulking when measuring by volume, the concrete/mortar may contain too little sand.

The term 'bulking' is also used when soil is excavated in construction projects and the volume of the excavated soil is more than its original volume when it was in a compact state. The bulking of soils is due to the loss of their compaction during the excavation process and may amount to about a 5–15% increase in volume. The bulking factor should be taken into account when working out the plant requirement for transporting the soil away from a site.

Example 7.15

A trench measuring 10.0 m × 0.6 m × 0.9 m deep is to be excavated in sandy clay. If the excavated soil bulks by 12%, find the volume of the soil after excavation.

Solution:

Volume of the soil in compact state = length × width × depth

= 10.0 × 0.6 × 0.9 (see also chapter 12)

= 5.4 m^3

Increase in the volume of the soil due to bulking $= \dfrac{12}{100} \times 5.4$

$= 0.648 \text{ m}^3$

Volume of the soil after bulking $= 5.4 + 0.648 = \mathbf{6.048 \text{ m}^3}$

Exercise 7.1

The solutions to Exercise 7.1 can be found in Appendix 2.

1. Express the shaded portions of the shapes shown in Figure 7.5 as a fraction.

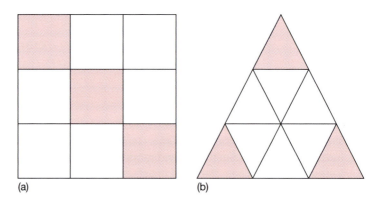

(a)   (b)

Figure 7.5

2. Express:
   (a) 20 pence as a fraction of £1.20
   (b) 250 grams as a fraction of 1.200 kg (1000 g = 1 kg)
   (c) 60° as a fraction of 120°.

3. Convert the following fractions into their lowest terms:
   (a) $\dfrac{10}{12}$;   (b) $\dfrac{12}{15}$;   (c) $\dfrac{40}{100}$.

4. Solve the following:
   (a) $\dfrac{2}{3} + \dfrac{5}{6}$;   (b) $\dfrac{1}{3} + \dfrac{1}{5}$;   (c) $\dfrac{3}{4} + \dfrac{2}{5}$;   (d) $\dfrac{3}{4} + \dfrac{4}{5}$.

5. Calculate:
   (a) $\dfrac{1}{3}$ of 33 m²;   (b) $\dfrac{1}{5}$ of £1.50;   (c) $\dfrac{3}{4}$ of 200 m.

6. Convert the following fractions into decimal numbers and percentages:
   (a) $\dfrac{3}{4}$;   (b) $\dfrac{3}{5}$;   (c) $\dfrac{7}{10}$;   (d) $\dfrac{4}{5}$.

7. Calculate: (a) 20% of 150; (b) 75% of 3000; (c) 90% of 900.

8. Nikki obtained 40 marks out of 60 in mathematics and 50 marks out of 80 in science. In which subject did she do better?

9. Ray wants to distribute part of £200 among his friends as shown:

   25% to Peter

   30% to Karen

   15% to Dave

   How much money will he be left with after distribution to the above?

10. A building society offers 4.5% annual interest for one of its accounts. Calculate the interest that Imran will get if he deposits £1200.00 for one year.

11. Kiran bought the following goods at a warehouse where the prices are displayed exclusive of VAT:

   One boiler: £845.00

   Seven radiators: £80.00 each

   Find the total amount paid by Kiran if VAT was charged at 20%.

12. A batch of concrete is made using the following materials:

   Cement = 200 kg

   Sand = 450 kg

   Gravel = 750 kg

   Water = 100 kg

   (a) Express the amount of sand as a fraction, and a percentage of the total mass of concrete.

   (b) Express the amount of water as a fraction, and a percentage of the amount of cement.

13. Two bricks were tested for water absorption. The results were:

Brick	Mass of dry brick (kg)	Mass of water absorbed (kg)
A	2.200	0.200
B	2.500	0.250

   If water absorption = $\dfrac{\text{mass of water absorbed}}{\text{mass of dry brick}}$, find the water absorption of each brick as a fraction and as a percentage.

14. The total heat loss from a building is 25 000 Watts. Find the heat loss through the windows and the roof, given that:

   • 15% of the total heat loss is through the windows

   • 35% of the total heat loss is through the roof.

15. 27 m² of external brickwork of a recently built building are affected by efflorescence. Express this as a fraction and as a percentage of 216 m², the total area of brickwork.

16. Figure 7.6 shows the amount of money spent on different construction activities in building a house extension. Express:

   (a) the money spent on foundations and brickwork as a fraction of the total cost

   (b) the money spent on finishes as a percentage of the total cost.

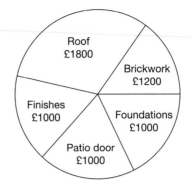

Figure 7.6

17. A sample of timber was tested in a laboratory to determine its moisture content. It was found that:

Mass of wet timber = 2.350 kg

Mass of dry timber = 2.0 kg

Find the moisture content of timber, as a percentage, if:

$$\text{Moisture content} = \frac{\text{Mass of wet timber} - \text{Mass of dry timber}}{\text{Mass of dry timber}}$$

Answers to Exercise 7.1

1. (a) $\frac{1}{3}$; (b) $\frac{1}{3}$

2. (a) $\frac{1}{6}$; (b) $\frac{5}{24}$; (c) $\frac{1}{2}$

3. (a) $\frac{5}{6}$; (b) $\frac{4}{5}$; (c) $\frac{2}{5}$

4. (a) $\frac{3}{2}$; (b) $\frac{8}{15}$; (c) $\frac{23}{20}$; (d) $\frac{31}{20}$

5. (a) 11 m^2 (b) £0.30   (c) 150 m

6. (a) 0.75/75%; (b) 0.6/60%; (c) 0.7/70%; (d) 0.8/80%;

7. (a) 30; (b) 2250; (c) 810

8. Mathematics (percentage marks: maths = 66.67; science = 62.5)

9. £60

10. £54

11. £1686.00

12. (a) $\frac{3}{10}$ /30%; (b) $\frac{1}{2}$ /50%

13. Brick A: $\frac{1}{11}$ /9.09%; Brick B: $\frac{1}{10}$ /10%

14. Windows: 3750 W; Roof : 8750 W

15. $\dfrac{1}{8}$ /12.5%

16. (a) $\dfrac{11}{30}$; (b) 16.67%

17. 17.5%

8

Graphs

Learning outcomes:

(a) Plot straight-line graphs from linear equations

(b) Plot experimental data and produce the 'best-fit' line

(c) Calculate the gradient of the straight line (m) and the intercept on the y-axis (c)

(d) Determine the law of the straight-line graph

8.1 Introduction

A graph shows the relationship between two variables. Graphical representation is a quick and easy way to show the information collected from a test or an observation. The shape of a graph depends on the data being plotted and helps us to draw conclusions. Figure 8.1 shows four typical graphs that we may come across in construction and engineering.

8.2 Cartesian axes and coordinates

A graph consists of two axes drawn at right angles to each other. The horizontal line is called the x-axis and the vertical line is called the y-axis. The point where the axes intersect is called the origin, and at this point the values of x and y are zero. The axes are known as rectangular or Cartesian axes, as shown in Figure 8.2.

The space on the right-hand side of the y-axis is used for the positive values of x. The space on the left-hand side of the y-axis is used for the negative values of x. The space above the x-axis is used to plot the positive values of y, and the space below for the negative values of y.

Each point that is to be plotted on a graph must have two values, i.e. the x value and the y value. These are also known as the x coordinate and the y coordinate, respectively. If the coordinates of a point are (3, 5) it means that:

- the first number, i.e. 3, is the x coordinate;
- the second number, i.e. 5, is the y coordinate.

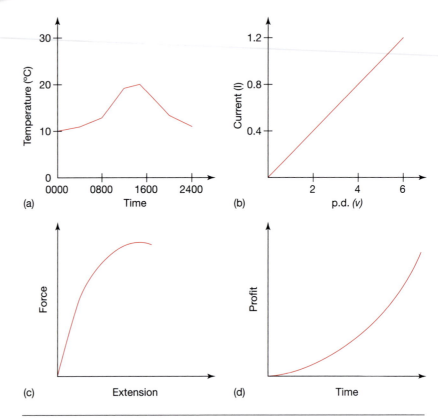

(a)

(b)

(c)

(d)

Figure 8.1

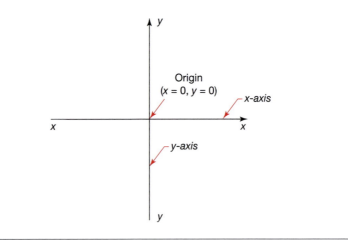

Figure 8.2

Graphs are usually plotted on A4-size graph paper. Appropriate scales are used for the axes so that all the data can be plotted on the graph paper. It is not necessary to use the same scale for the two axes, but whatever scales are selected they should be easy to use.

Example 8.1

Plot the following points on a graph: A (–5, –3), B (–6, 5), C (3, –7), D (8, 6)

Solution:

Assume that each centimetre on the graph paper represents two units, as shown in Figure 8.3.

Point A has $x = -5$ and $y = -3$

A point is located on the x-axis where $x = -5$. From this point, a vertical line is drawn downwards as the value of y is negative. A horizontal line is drawn through the –3 mark on the y-axis. The point where these two lines meet is the plot of point A.

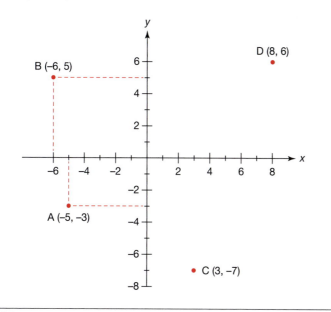

Figure 8.3

Point B has $x = -6$ and $y = 5$. The procedure, as before, involves the location of the point on the x-axis where $x = -6$. From this point, a vertical line is drawn upwards as the value of y is positive. A horizontal line is drawn through the 5 mark on the y-axis. The point where these two lines meet is the plot of point B. Following this procedure, points C and D can be plotted as well, as shown in Figure 8.3.

Example 8.2

The air temperatures on last year's hottest day were:

Time	0800	1000	1200	1400	1600	1800	2000
Temperature (°C)	16	20	28	30	29	27	24

Plot a graph to show the air temperature against time.

Solution:

On graph paper the axes are drawn as shown in Figure 8.4. The x-axis is normally used for the independent variable and the y-axis for the dependent variable. Time is the independent variable and is marked on the x-axis. Temperature, the dependent variable, is marked on the y-axis as shown in Figure 8.4. As 16°C is the lowest temperature, the first point on the y-axis can be anything from 0°C to 14°C. But here we will start from 12°C. The first point on the x-axis is 0800. The scales used in this example are:

- x-axis: 1 cm = 2 hours
- y-axis: 1 cm = 4°C

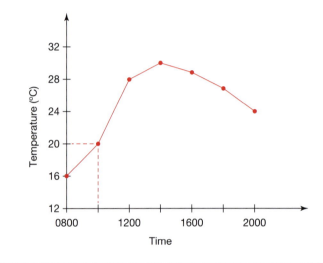

Figure 8.4

The air temperature at 0800 hours was 16°C; to mark this point on the graph paper we must draw a vertical line through the 0800 hours mark and a horizontal line through the 16°C mark. As the vertical line through the 0800 mark is the y-axis and already drawn, mark a point at the 16°C mark with a cross or a dot.

At 1000 hours the temperature was 20°C. To mark this point on the graph, draw a vertical line through the 1000 hours mark and a horizontal line through the 20°C mark. Mark the point where the two lines meet with a dot or a cross. Repeat this process to mark the other points and join them, as shown in Figure 8.4

Example 8.3

A material was heated to 68°C and then allowed to cool down to room temperature. Its temperature, as it cooled, was recorded and is given below:

Time (minutes)	0	0.5	1	2	3	4	6	8	10	15
Temperature (°C)	68	65	63	60	57	55	52	50	48	46

Produce a graph of temperature versus time by drawing a smooth curve through all points.

Solution:

On a graph paper draw the axes as shown in Figure 8.5. The scales used for the axes are:

- *x*-axis: 1 cm = 2 minutes
- *y*-axis: 1 cm = 4°C

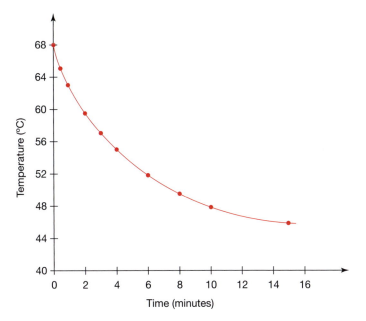

Figure 8.5

The time and temperature are marked on the axes as shown in Figure 8.5. The coordinates of the points to be plotted are: (0, 68), (0.5, 65), (1, 63), (2, 60), (3, 57), (4, 55), (6, 52), (8, 50), (10, 48) and (15, 46). The points are plotted as described in the earlier examples and joined by a smooth curve.

8.3 Straight-line graphs

The general equation of the straight line is given by:

$y = mx + c,$

where *x* is the independent variable and *y* the dependent variable, *m* is the slope or gradient of the straight line and *c* is the intercept made by the straight line on the *y*-axis.

Any equation that is written in this form will produce a straight-line graph, as explained in Examples 8.4 and 8.5.

Example 8.4

Draw the graph of $y = 2x + 3$ from $x = -2$ to $x = 4$

Solution:

Before plotting the points on the graph paper we need to use different values of x from the given range, and find the corresponding values of y.

If $x = -2$, $y = (2 \times -2) + 3$

$\quad\quad = -4 + 3 = -1$

Similarly for $x = -1, 1, 2$ and 4, the corresponding values of y are determined, which are:

x	-2	-1	1	2	4
y	-1	1	5	7	11

We now have five points, the coordinates of which are: $(-2, -1)$, $(-1, 1)$, $(1, 5)$, $(2, 7)$ and $(4, 11)$. Suitable scales are selected and the points plotted, as shown in Figure 8.6. A straight line is drawn that passes through all the points. This is the graph of the equation: **$y = 2x + 3$**. (Actually, three points should be sufficient to produce a straight-line graph).

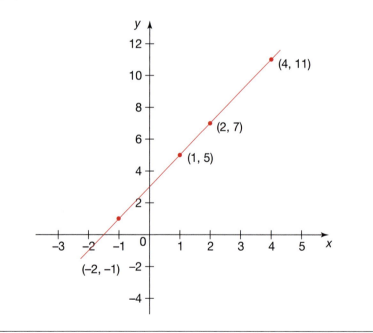

Figure 8.6

Example 8.5

Draw the graph of $y = 5 - 6x$, between the limits $x = -2$ and $x = 3$.

Solution:

Use different values of x, from the given range, and determine the corresponding values of y.

If $x = -2$, $y = 5 - (6 \times -2)$

$\qquad = 5 + 12 = 17$

Similarly for $x = -1, 1, 2$ and 3, the corresponding values of y are calculated. These are:

x	−2	−1	1	2	3
y	17	11	−1	−7	−13

The coordinates of the points are: (−2, 17), (−1, 11), (1, −1), (2, −7) and (3, −13). The points are plotted as shown in Figure 8.7, and a straight line drawn passing through all the points. This is the graph of the equation:
y = 5 − 6x

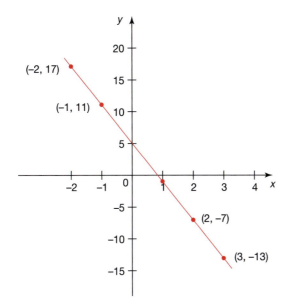

Figure 8.7

8.4 The law of the straight line

In building technology and civil engineering, practical work is often undertaken to understand and prove some fundamental concepts. In some cases the results obtained from experimental work produce a perfect straight-line graph or linear relationship, whereas in others the points do not exactly lie on a straight line but exhibit a linear relationship. A best-fit line is drawn in the latter case. The law of the straight line can be determined using the equation $y = mx + c$.

8.4.1 The gradient (*m*)

The gradient (*m*) of a straight line is a measure of how steep the line is. To calculate the gradient, draw a right-angled triangle of a reasonable size on the straight line. Find the length of the vertical and the horizontal sides of the triangle using the scales of the axes. Figure 8.8 shows a straight-line graph and a right-angled triangle ABC.

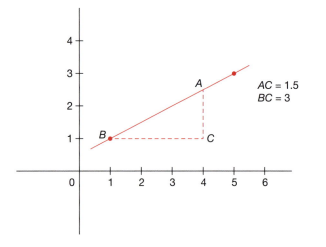

Figure 8.8

$$\text{Gradient of the line, } m = \frac{AC}{BC} = \frac{1.5}{3} = 0.5$$

The gradient is positive if the line rises, i.e. the right end of the line is higher than the left end. The gradient is negative if the left end of the line is higher than the right end, as shown in Figure 8.9.

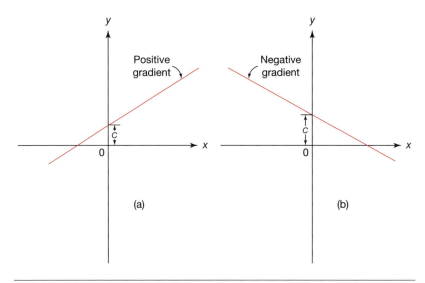

Figure 8.9

8.4.2 The intercept (c)

The intercept made by the straight-line graph on the y-axis, when $x = 0$, is called the y-axis intercept (see Figure 8.9). If x is not equal to zero at the origin, then the coordinates of a point that lies on the line are noted and substituted in the equation $y = mx + c$, to obtain the value of c.

Example 8.6

Find the law of the straight line shown in Figure 8.10

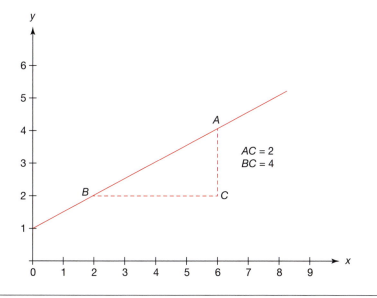

Figure 8.10

Solution:

The intercept on the y-axis (c) = 1. To find the gradient of the straight line, a right-angled triangle ABC is drawn as shown in Figure 8.10. The right end of the line is higher than the left end – therefore the gradient is positive.

Gradient $m = \dfrac{AC}{BC} = \dfrac{2}{4} = 0.5$

As $y = mx + c$

∴ The law of the straight line is: $y = 0.5x + 1$

Example 8.7

Find the law of the straight line shown in Figure 8.11

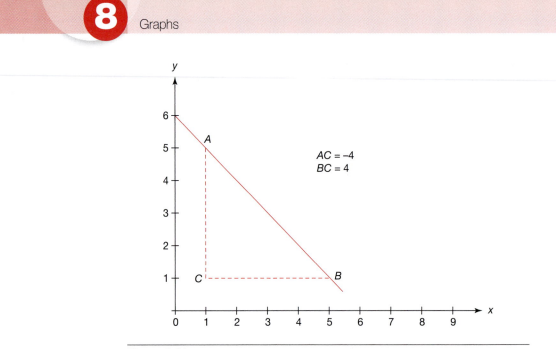

Figure 8.11

Solution:

The intercept on the y-axis (c) = 6. To find the gradient, a right-angled triangle ABC is drawn as shown in Figure 8.11. As the left end of the line is higher than the right end, the gradient of the line is negative.

$$\text{Gradient } m = \frac{AC}{BC} = -\frac{4}{4} = -1$$

∴ The law of the straight line is: $y = -1x + 6$ ($y = mx + c$)

or **$y = -x + 6$**

Example 8.8

The results of a tensile test on a metal are:

Stress (S)	0	0.04	0.08	0.10	0.12	0.16 kN/m^2
Strain (N)	0	0.0004	0.0008	0.001	0.0012	0.0016

(a) Select suitable scales and plot a graph of stress versus strain, and show that the equation connecting stress (S) and strain (N) is of the form $S = mN + c$

(b) Find the equation of the graph

(c) Find the stress in the metal when the strain is increased to 0.002 (assuming that the behaviour of the metal does not change)

Solution:

(a) In this example the independent and the dependent variables are different from those used earlier. N is used instead of x, and S is used instead of y. Stress (S) is plotted on the y-axis and strain (N) on the x-axis.

Suitable scales are selected and marked on the axes as shown in Figure 8.12. It is evident on plotting the points that they follow a straight line. Therefore we can say that the equation connecting S and N is of the form $S = mN + c$.

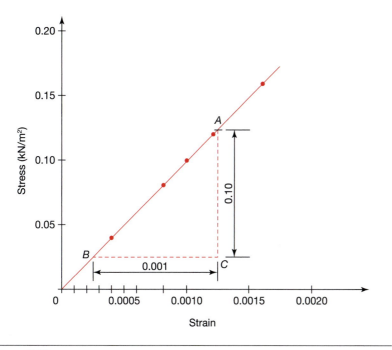

Figure 8.12

(b) A straight line is drawn through the points and the gradient m and intercept c are determined as explained in Example 8.6.

Gradient $m = \dfrac{AC}{BC} = \dfrac{0.10}{0.001} = 100$

Intercept $c = 0$

The equation of the straight line is $S = mN + c$

$$= 100N + 0$$

$$\text{or } \mathbf{S = 100N}$$

(c) When $N = 0.002$, $S = 100 \times 0.002 = \mathbf{0.2\ kN/m^2}$

8.5 Simultaneous equations

The procedure for drawing the graphs of straight lines has already been explained in Section 8.3. The simultaneous equations should be rearranged so that their form is similar to the standard straight-line equation. The graphs are plotted on the same axes and the answer obtained by finding the coordinates of the point of intersection. The point of intersection is common to both lines (or equations), as shown in Example 8.9.

Example 8.9

Solve graphically the simultaneous equations:

$$x + 3 = -2y$$

and

$$x - y = 3$$

Solution:

The equations may be given in any form, but it is necessary to rearrange each equation into the form of the straight-line law, i.e. $y = mx + c$. The first equation can be written as:

$$2y = -x - 3$$

$$\text{or } y = \frac{-x - 3}{2} \tag{1}$$

Similarly, the second equation becomes:

$$y = x - 3 \tag{2}$$

At least three values of x are assumed in each case and the corresponding values of y determined:

<u>Equation 1</u>: If $x = 1$, $y = \dfrac{-1 - 3}{2} = \dfrac{-4}{2} = -2$

If $x = 3$, $y = \dfrac{-3 - 3}{2} = \dfrac{-6}{2} = -3$

Similarly when $x = 5$, $y = -4$

The coordinates for Equation (2) are calculated in a similar manner. The following table summarises the calculations:

Equation 1	x	1	3	5
	$y = \dfrac{-x - 3}{2}$	−2	−3	−4
Equation 2	x	1	3	5
	$y = x - 3$	−2	0	2

The graphs are plotted as shown in Figure 8.13. The solution of the two equations is the point of intersection (1, −2). Therefore **$x = 1$ and $y = -2$**.

8.6 Quadratic equations

The graphical method, although time consuming, can be used to solve any quadratic equation. There is only one unknown (e.g. x) in a quadratic equation, therefore for the graphical solution it is necessary to equate the quadratic equation to y. For example, if we are asked to solve the equation, $2x^2 - 6x - 8 = 0$, then as a starting point we say that:

$$y = 2x^2 - 6x - 8 = 0$$

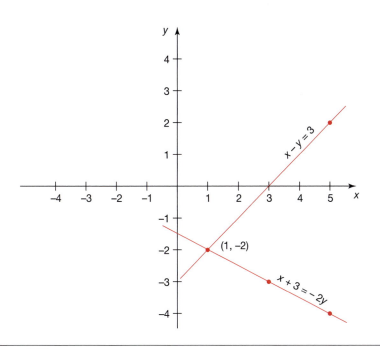

Figure 8.13

At least six values of x are assumed and the corresponding values of y are determined. The points are plotted on a graph and joined by a smooth curve. The solution of the equation is determined by finding the values of x coordinates where the curve meets the x-axis ($y = 0$ on the x-axis).

Example 8.10

Solve the equation $2x^2 - 6x - 8 = 0$, by graphical method.

Solution:

Initially five values of x (−4, −2, 0, 2, 4) are assumed and the corresponding values of y calculated as shown below. Depending on the shape of the graph, further points may be necessary to find the solution.

x	−4	−2	0	2	4
$2x^2$	32	8	0	8	32
$-6x$	24	12	0	−12	−24
−8	−8	−8	−8	−8	−8
$2x^2 - 6x - 8$	48	12	−8	−12	0

The equation $2x^2 - 6x - 8$ is evaluated by adding −8 and the values of $2x^2$ and $-6x$. This is also the value of the y coordinate, as $y = 2x^2 - 6x - 8$.

We have five points (−4, 48), (−2, 12), (0, −8), (2, −12) and (4, 0), which are plotted as shown in Figure 8.14. A smooth curve is drawn passing through all the points.

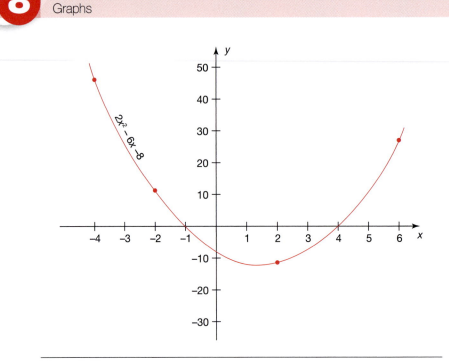

Figure 8.14

As the last point just touches the x-axis, another point is needed to extend the curve. The value of y is determined at $x = 6$, as shown:

x	−4	−2	0	2	4	6
$2x^2$	32	8	0	8	32	72
$-6x$	24	12	0	−12	−24	−36
-8	−8	−8	−8	−8	−8	−8
$2x^2 - 6x - 8$	48	12	−8	−12	0	28

The curve crosses the x-axis at $x = -1$ and $x = 4$, therefore the solutions of the equation $2x^2 - 6x - 8 = 0$ are:

$x = -1$ and $x = 4$

Exercise 8.1

The solutions to Exercise 8.1 can be found in Appendix 2.

1. The figures below show the profit made by a building company during the last five years. Represent the data in the form of a graph.

Year	1	2	3	4	5
Profit (£)	60 000	100 000	200 000	160 000	230 000

2. A test was conducted to determine the effect of water content on the density of compacted soil. The results are shown below:

Density (kg/m³)	1810	1860	1880	1885	1830
Water content (%)	10	12	13	14	16

Draw a graph between density and water content, showing density on the y-axis.

3. Draw the graphs of the following equations:
   (a) $y = 2x - 4$, taking values of x between -2 and 4.
   (b) $y = -0.6x + 5$, taking values of x between -2 and 4.

4. The figures below show how the current through a metal wire varies with the potential difference applied across its ends:

Potential difference (V)	0.65	1.26	1.90	2.50	3.15
Current (I)	0.1	0.2	0.3	0.4	0.5

   (a) Plot a graph of potential difference (y-axis) against current (x-axis).
   (b) Find the gradient of the graph and determine the law connecting V and I.

5. In an experiment to prove Hooke's Law, the following data were obtained by stretching a spring:

Stretching force, F (Newton)	0	1	2	3	4	5	6
Extension, L (mm)	0	8.5	16	24.5	33	41.5	50

   (a) Plot a graph of stretching force (y-axis) against extension (x-axis) and find the gradient of the straight line.
   (b) Show that the law connecting F and L is of the form $F = mL$, and find the law.

6. The results of a tensile test on steel are shown below:

Stress (S) kN/m²	0	0.08	0.16	0.2	0.24	0.32
Strain (N)	0	0.0004	0.0008	0.0010	0.0012	0.0016

   (a) Plot a graph of stress versus strain. Show that the law connecting stress (S) and strain (N) is of the form $S = mN$, and find the law.
   (b) Find the stress in steel when the strain is increased to 0.002 (assuming that the straight-line law holds good outside the range of the given data).

7. A sample of plastic pipe expanded as shown below:

Length, L (mm)	1000	1001.1	1002.3	1003.1
Temperature, T (°C)	20	32	45	54

   (a) Plot a graph of expansion (E) against temperature (T) and find the equation of the line.
   (b) Find the length of the pipe at 10°C.

8. Figure 8.15 shows the relationship between the compressive strength and the water content of concrete. Find the equation of the straight line.

9. Solve the following simultaneous equations graphically:
   $x + y = 5$, and $3x - 2y = -5$

10. Solve graphically the quadratic equation: $x^2 + x - 20 = 0$.

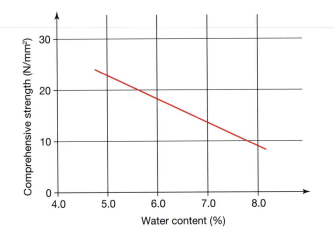

Figure 8.15

Answers to Exercise 8.1

1. See Appendix 2

2. See Appendix 2

3. See Appendix 2

4. (a) See Appendix 2

   (b) $m = 6.25$; $c = 0$; Law: $V = 6.25I$

5. (a) See Appendix 2

   (b) $m = 0.12$; $c = 0$; Law: $F = 0.12L$

6. (a) See Appendix 2

   $m = 200$ and $c = 0$

   $S = 200N$

   (b) $S = 0.4$ kN/m^2

7. (a) See Appendix 2 (consider 20°C as the initial temperature)

   $m = 0.091$; $c = -1.8$; Law: $E = 0.091T - 1.8$

   (b) 999.11 mm

8. $m = -4.583$; $c = 45.67$; Law: $Y = -4.583x + 45.67$

9. $x = 1$, $y = 4$ (see graph in Appendix 2)

10. $x = -5$, $x = 4$ (see graph in Appendix 2)

CHAPTER **9**

Units and their conversion

Learning outcomes:

(a) Use graphs and conversion factors to convert the units of length, mass, area and volume from one system to the other

(b) Use tables to convert the units of temperature

9.1 Introduction

Units of measurement, in one form or another, have been with us for many centuries. It is quite likely that the units for length and mass were the first ones to be invented. Our ancestors used the body parts as measuring instruments. Early Babylonian, Egyptian and other records indicate that length was measured with the forearm, hand or fingers. As civilisations evolved, units of measurement became more complicated to cater for trade, land division, taxation and other uses. As the means of communication were not very good before the eighteenth century, different units were developed in different countries, or even different parts of the same country, for the same purpose. In Britain, units from Egyptian, Babylonian, Anglo-Saxon and other European cultures evolved into the inch, foot and yard. In Britain and many other countries, the FPS system had been in use till the late 1950s. The basic units for length, mass and time were the foot, pound and second.

During the 1960s scientists and mathematicians started to use the metric system to simplify their calculations and promote communication across different nations. Even in the metric system two branches of systems co-existed for a long time: the CGS and MKS systems. The base units for length, mass and time, in the two systems are:

CGS system   the centimetre, gram and second

MKS system   the metre, kilogram and second.

In 1960 the metric system was revised and simplified for international use. The metre, kilogram and second were kept as the basic units for length, mass and time. This system, which includes four other basic units, is called

Table 9.1

	CGS/MKS system	FPS system
Length	Millimetre (mm), centimetre (cm), metre (m), kilometre (km)	Inch (in), foot (ft), yard (yd), mile
Mass	Gram (g), kilogram (kg), tonne	Ounce (oz), pound (lb), stone, ton
Time	Second (s)	Second (s)
Temperature	Degrees Celsius (°C)	Degrees Fahrenheit (°F)
Capacity	Millilitre (ml), centilitre (cl), litre (l)	Pint, gallon
Volume	Cubic millimetre, cubic centimetre, cubic metre	Cubic foot, cubic yard

the SI system (International System of Units). A brief comparison of the metric systems and the FPS system is given in Table 9.1.

Although in formal work the metric units have replaced the British units, the inch, mile, ounce, pound, stone, pint, gallon, etc. are still used in everyday life in Britain, and it becomes necessary sometimes to convert a unit from one system into the other.

Conversion of units may be done using a range of methods:

- conversion factors
- tables
- graphs

The following sections give examples of converting the units of length, mass, temperature, etc. based on these methods.

9.2 Length

The SI unit of length is metre, but millimetre, centimetre and kilometre are quite commonly used in mathematics. The use of conversion factors gives the most accurate results, but the use of graphs is a fairly straightforward method.

9.2.1 Conversion factors

Table 9.2 gives the conversion factors for some of the commonly used units in the metric as well as the FPS system:

Example 9.1

Convert: (a) 56 mm into metres

(b) 445 mm into metres

Table 9.2

1000 micrometres = 1 mm	1 cm = 10 mm	1 in = 25.4 mm
1 mm = 0.1 cm	1 mm = 0.03937 in	1 in = 2.54 cm
1 mm = 0.001 m	1 mm = 0.003281 feet	1 ft = 304.8 mm
1 cm = 0.01 m	1 cm = 0.3937 in	1 ft = 30.48 cm
1 m = 0.001 km	1 cm = 0.03281 ft	1 ft = 0.3048 m
1 km = 1000 m	1 m = 3.281 ft	1 ft = 12 in
1 m = 1000 mm	1 m = 1.0936 yd	1 yd = 0.9144 m
1 m = 100 cm	1 in = 0.0254 m	1 mile = 1.609 344 km

(c) 5.4 m into centimetres

(d) 2 ft 5 in into millimetres

(e) 4 yd into metres

(f) 10 in into metres

Solution:

(a) To convert millimetres into metres use the conversion factor that has millimetre on the left and metre on the right, i.e. 1 mm = 0.001 m. This technique will be used for other conversions as well. Therefore:

If 1 mm = 0.001 m, 56 mm will be equal to 56 times 0.001

56 mm = 56 × 0.001 = **0.056 m**

(b)

1 mm = 0.001 m

445 mm = 445 × 0.001 = **0.445 m**

(c)

1 m = 100 cm

5.4 m = 5.4 × 100 = **540 cm**

(d) Convert 2 ft 5 in into inches first, then into millimetres:

1 ft = 12 inches

2 ft = 2 × 12 = 24 in

2 ft 5 in = 24 + 5 in = 29 in

Now convert inches into millimetres using 1 in = 25.4 mm:

29 in = 29 × 25.4 = **736.6 mm**

(e)

1 yd = 0.9144 m

4 yd = 4 × 0.9144 = **3.6576 m**

(f)

$$1 \text{ in} = 0.0254 \text{ m}$$
$$10 \text{ in} = 10 \times 0.0254 = \textbf{0.254 m}$$

9.2.2 Use of the graphical method

A graph can be plotted for any two units, and the resulting straight line used to convert one unit to the other. Figure 9.1 shows a graph that can be used to convert metres into feet and vice versa. Known conversion factors are used to plot the straight-line graph:

$$1 \text{ ft} = 0.3048 \text{ m}$$
$$10 \text{ ft} = 3.048 \text{ m}$$
$$20 \text{ ft} = 6.096 \text{ m}$$
$$30 \text{ ft} = 9.144 \text{ m}$$
$$40 \text{ ft} = 12.192 \text{ m}$$

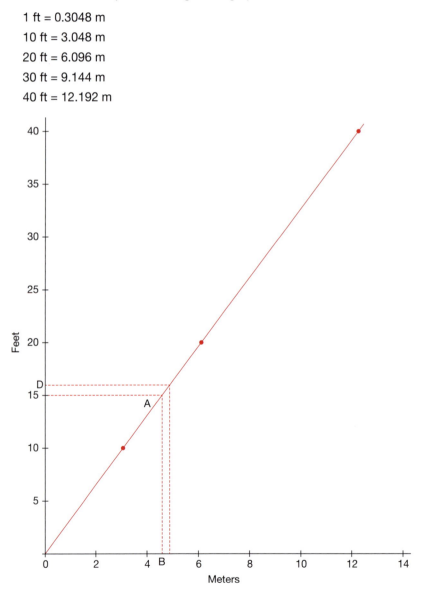

Figure 9.1

In Figure 9.1 the x-axis is used for metres and y-axis for feet; however, the choice of an axis for a particular unit is immaterial. Using the scale shown in Figure 9.1, plot four points having the coordinates (3.048, 10), (6.096, 20), (9.144, 30) and (12.192, 40). Draw a straight line through all the points. This graph can be used to convert feet into metres and vice versa, as shown in Example 9.2.

Example 9.2

Draw a graph between feet and metres and use it to convert:

(a) 15 ft into metres

(b) 5 m into feet

Solution:

The procedure for drawing the graph has already been explained above. The following method explains how to use this graph for the conversion of units:

1. Draw a horizontal line from the 15 feet mark and extend it to meet the straight-line graph at point A (Figure 9.1). Draw a vertical line from point A to meet the x-axis at point B. Point B represents **4.51 m**, which is the required conversion.

2. Draw a vertical line from the 5 m mark to meet the straight-line graph at point C. From point C draw a horizontal line to meet the y-axis at point D. Read off point D to get the conversion. The answer is **16.5 feet**.

9.3 Mass

The SI unit of mass is kilogram, but gram and tonne are also used in mathematics, science and daily use. Two methods are given in the following sections to convert from one system into the other.

9.3.1 Conversion factors

Table 9.3 gives the conversion factors for some of the commonly used units in the metric as well as the FPS system:

Table 9.3

1 g = 0.001 kg	16 oz = 1 lb	1 g = 0.03527 oz
1 g = 0.000001 tonne	14 lb = 1 stone	1 kg = 2.204 62 lbs
1 kg = 1000 g	112 lb = 1 hundred weight (cwt)	1 lb = 453.59237 g
1 kg = 0.001 tonne	20 cwt = 1 ton	1 lb = 0.45359 kg
1 tonne = 1000 kg	2240 lb = 1 ton	1 oz = 28.3495 g
1 tonne = 1 000 000 g		

Example 9.3

Convert: (a) 0.050 kg into grams
(b) 20 500 kg into tonnes
(c) 505 g into ounces

Solution:

(a) To convert kilograms into grams, use the conversion factor that has kilogram on the left and gram on the right, i.e. 1 kg = 1000 g:

1 kg = 1000 g.

0.050 kg = 0.050 × 1000 = **50 g**

(b)

1 kg = 0.001 tonne

20 500 kg = 20 500 × 0.001 = **20.5 tonne**

(c)

1 g = 0.03527 oz

505 g = 505 × 0.03527 = **17.811 oz**

9.3.2 Graphical method

As explained in Section 9.2.2, a straight-line graph can be plotted between two units and used to convert one unit into the other. Figure 9.2 shows a graph that can be used to convert kilograms into pounds and vice versa. Known conversion factors are used to plot the straight-line graph:

1 kg = 2.20462 lb

10 kg = 22.0462 lb

20 kg = 44.0924 lb

30 kg = 66.1386 lb

In Figure 9.2, the x-axis represents kilograms and the y-axis represents pounds. Use the scales as shown in the figure and draw a straight line passing through all the points. Using this straight line will result into the desired conversion, i.e. kilograms into pounds or pounds into kilograms. Example 9.4 explains this process.

Example 9.4

Use two methods to convert the following units and compare the results:

(a) Convert 10.500 kg into pounds

(b) Convert 50 lb into kilograms

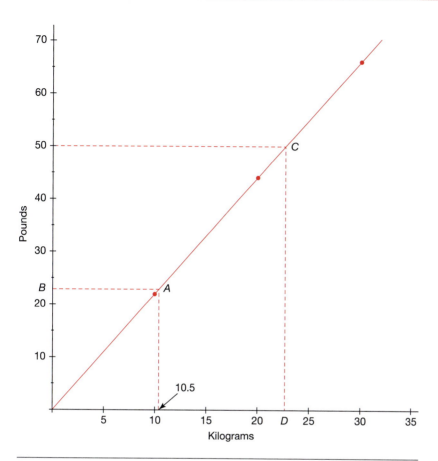

Figure 9.2

Solution:

(a) Graphical method. Use the graph between kilograms and pounds as shown in Figure 9.2. To convert 10.5 kg into pounds, draw a vertical line from the 10.5 kg mark on the *x*-axis to meet the straight-line graph at point A. Draw a horizontal line from point A to meet the *y*-axis at point B. Read off point B to get the answer in pounds.

Answer = **23 lb**

Use of conversion factors. 1 kg = 2.20462 lb

10.5 kg = 10.5 × 2.204 62 = **23.15 lb**

(b) Graphical method. To convert 50 lb into kilograms, use the procedure already explained in the first part of this example, but start from 50 lb on the *y*-axis. Draw a horizontal line to meet the straight-line graph at point C. From point C draw a vertical line to meet the *x*-axis at point D. Read off point D to get the answer.

Answer = **22.75 kg**

Use of conversion factors: 1 lb = 0.4536 kg

50 lb = 50 × 0.4536 = **22.68 kg**

A comparison of the two methods shows that the results from the two methods are slightly different:

	Graph	Conversion factor
10.500 kg	23.0 lb	23.15 lb
50 lb	22.75 kg	22.68 kg

The graphical method produces quick but approximate answers.

9.4 Area, volume and capacity

For converting the units of area, volume and capacity, a small selection of conversion factors are given in Table 9.4.

Table 9.4

Area	Volume	Capacity
$1 \text{ mm}^2 = 0.01 \text{ cm}^2$	$1 \text{ mm}^3 = 0.001 \text{ cm}^3$	1000 millilitres (ml) = 1 litre
$1 \text{ mm}^2 = 0.000001 \text{ m}^2$	$1 \text{ mm}^3 = 1 \times 10^{-9} \text{ m}^3$	100 centilitres (cl) = 1 litre
$1 \text{ cm}^2 = 0.0001 \text{ m}^2$	$1 \text{ cm}^3 = 1 \times 10^{-6} \text{ m}^3$	1 ml = 0.001 litre
$1 \text{ cm}^2 = 100 \text{ mm}^2$	$1 \text{ cm}^3 = 1000 \text{ mm}^3$	1 cl = 0.01 litre
$1 \text{ m}^2 = 1\,000\,000 \text{ mm}^2$ or $1 \times 10^6 \text{ mm}^2$	$1 \text{ m}^3 = 1 \times 10^9 \text{ mm}^3$	1 litre = 1000 cm^3
$1 \text{ m}^2 = 10\,000 \text{ cm}^2$	$1 \text{ m}^3 = 1 \times 10^6 \text{ cm}^3$	1000 litre = 1 m^3
		1 pint = 0.5683 litre
		1 gallon = 4.5461 litre
		1 litre = 0.22 gallon

Example 9.5

Convert: (a) 15 604 mm^2 into cm^2
(b) 256 000 mm^2 into m^2
(c) 3.56 m^2 into mm^2

Solution:

(a) Use the conversion factor that involves mm^2 and cm^2:

$1 \text{ mm}^2 = 0.01 \text{ cm}^2$

$15\,604 \text{ mm}^2 = 15\,604 \times 0.01 = \textbf{156.04 cm}^2$

(b) This question involves mm² and m²:

1 mm² = 0.000001 m²

256 000 mm² = 256 000 × 0.000001 = **0.256 m²**

(c)

1 m² = 10⁶ mm²

3.56 m² = 3.56 × 10⁶ = **3 560 000 mm²**

Example 9.6

Convert: (a) 6 593 000 mm³ into m³

(b) 0.00024 m³ into mm³

Solution:

(a)

1 mm³ = 1 × 10⁻⁹ m³

6 593 000 mm³ = 6 593 000 × 1 × 10⁻⁹ = **0.0066 m³**

(b)

1 m³ = 1 × 10⁹ mm³

0.00024 m³ = 0.00024 × 1 × 10⁹ = **240 000 mm³**

Example 9.7

Convert: (a) 12 560 ml into litres

(b) 6.67 litres into centilitres

(c) 0.0053 m³ into litres

(d) 30 litres into gallons

Solution:

(a)

1 ml = 0.001 litre

12 560 ml = 12 560 × 0.001 = **12.560 litres**

(b)

1 litre = 100 cl

6.67 litres = 6.67 × 100 = **667 cl**

(c)

1 m³ = 1000 litres

0.0053 m³ = 0.0053 × 1000 = **5.3 litres**

(d)

1 litre = 0.22 gallons

30 litres = 30 × 0.22 = **6.6 gallons**

9.5 Temperature

Temperature may be converted using a range of methods, but here only the use of tables is given. Table 9.5 shows a selection of temperatures in degrees Celsius and their equivalent temperatures in degrees Fahrenheit. It shows that for a change of 1°C the corresponding change in degrees Fahrenheit is 1.8.

Table 9.5

Temperature		Temperature		Temperature	
°C	°F	°C	°F	°C	°F
0	32	40	104	80	176
5	41	45	113	85	185
10	50	50	122	90	194
15	59	55	131	95	203
20	68	60	140	100	212
25	77	65	149	105	221
30	86	70	158	110	230
35	95	75	167		

Example 9.8

Convert the following temperatures into degrees Fahrenheit:
(a) 21°C
(b) 37.5°C
(c) 74°C

Solution:

(a) A change of 1°C = a change of 1.8°F:

21°C = 20°C + 1°C

From Table 9.5, 20°C = 68°F

∴ 21°C = 68 + 1.8 = **69.8°F**

(b) 37.5 is an average of 35 and 40:

35°C = 95°F and 40°C = 104°F

The average of 95 and 104 is 99.5

∴ 37.5°C = **99.5°F**

(c)

74°C = 75°C − 1°C

= 167°F − 1.8°F

= **165.2°F**

Exercise 9.1

The solutions to Exercise 9.1 can be found in Appendix 2.

1. Convert: (a) 86 mm into metres

    (b) 385 mm into metres

    (c) 7.4 m into centimetres

    (d) 3 ft 2 in into millimetres

    (e) 6 yd into metres

    (f) 14 in into metres.

2. Use two methods to convert the following units and compare the results:

    (a) 24 ft into metres

    (b) 9 m into feet.

3. Convert: (a) 0.070 kg into grams

    (b) 25 500 kg into tonnes

    (c) 815 g into ounces.

4. Convert: (a) 15.500 kg into pounds

    (b) 250 kg into pounds

    (c) 75 lb into kilograms.

5. Convert: (a) 12 550 mm^2 into cm^2

    (b) 655 000 mm^2 into m^2

    (c) 2.75 m^2 into mm^2

    (d) 5.25 m^2 into cm^2.

6. Convert: (a) 95 450 mm^3 into m^3

    (b) 9545 mm^3 into cm^3

    (c) 0.00094 m^3 into mm^3.

7. Convert: (a) 12 560 ml into litres

    (b) 6.67 litres into centilitres

    (c) 350 cl into litres

    (d) 0.0053 m^3 into litres

    (e) 25 litres into gallons.

8. Convert 37°C and 61.5°C to degrees Fahrenheit.

Answers to Exercise 9.1

1. (a) 0.086 m; (b) 0.385 m; (c) 740 cm; (d) 965.2 mm; (e) 5.4864 m;
    (f) 0.3556 m

2. (a) 7.25 m (using graph); 7.315 (using conversion factor)

    (b) 29 feet 9 inches (using graph); 29 feet 6.3 inches (using conversion factors)

3. (a) 70 g; (b) 25.5 tonnes; (c) 28.745 oz
4. (a) 34.172 lb; (b) 551.155 lb; (c) 34.019 kg
5. (a) 125.5 cm^2; (b) 0.655 m^2; (c) 2 750 000 mm^2; (d) 52 500 cm^2
6. (a) 9.545 × 10^{-5} m^3; (b) 9.545 cm^3; (c) 940 000 mm^3
7. (a) 12.560 litres; (b) 667 cl; (c) 3.5 litres; (d) 5.3 litres; (e) 5.5 gallons
8. 98.6°F and 142.7°F

CHAPTER 10

Geometry

Learning outcomes:

(a) Identify the different types of angles, triangles and quadrilaterals

(b) Find angles in triangles, quadrilaterals and other geometrical constructions

(c) Use Pythagoras' theorem to determine diagonals in quadrilaterals and sides of right-angled triangles

(d) Calculate the circumference of a circle

10.1 Angles

When two straight lines meet at a point an angle is formed, as shown in Figure 10.1. There are two ways in which an angle can be denoted, i.e. either $\angle CAB$ or $\angle A$.

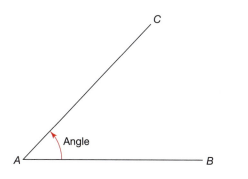

Figure 10.1

The size of an angle depends on the amount of rotation between two straight lines, as illustrated in Figure 10.2. Angles are usually measured in degrees, but they can also be measured in radians. A degree, defined as $\dfrac{1}{360}$ of a complete revolution, is easier to understand and use as compared to the radian. Figure 10.2 shows that the rotation of line AB makes (a) $\frac{1}{4}$ revolution or 90°, (b) $\frac{1}{2}$ revolution or 180°, (c) $\frac{3}{4}$ revolution or 270° and (d) a complete revolution or 360°.

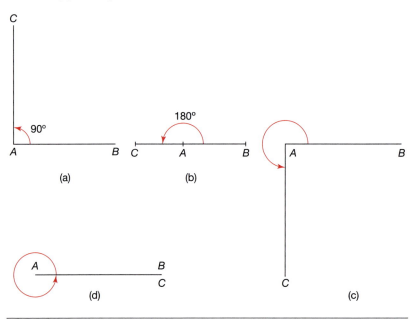

Figure 10.2

For accurate measurement of an angle a degree is further divided into minutes and seconds. There are 60 minutes in a degree and 60 seconds in a minute. This method is known as the sexagesimal system:

60 minutes (60′) = 1 degree

60 seconds (60″) = 1 minute (1′)

The radian is also used as a unit for measuring angles. The following conversion factors may be used to convert degrees into radians and vice versa.

1 radian = 57.30° (correct to 2 d.p.)

π radians = 180° (π = 3.14159; correct to 5 d.p.)

2π radians = 360°

Example 10.1

Convert: (a) 20°15′25″ into degrees (decimal measure)

(b) 32.66° into degrees, minutes and seconds.

(c) 60°25′45″ into radians.

Solution:

(a) The conversion of 15′25″ into degree involves two steps. The first step is to change 15′25″ into seconds, and the second to convert seconds into a degree. This is added to 20° to get the final answer.

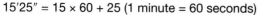
$15'25'' = 15 \times 60 + 25$ (1 minute = 60 seconds)

$= 900 + 25 = 925$ seconds

$= \dfrac{925}{3600}$ degree $= 0.257°$

$20°15'25'' = 20 + 0.257 = \mathbf{20.257°}$

(b) In 32.66°, 32° need no conversion but 0.66° will be converted into minutes and seconds.

$0.66° = 0.66 \times 60$ minutes $= 39.6'$

$0.6' = 0.6 \times 60$ seconds $= 36''$

Hence 32.66° = **32°39'36''**

(c)

$60°25'45'' = 60.4292°$

1 degree $= \dfrac{\pi}{180}$ radians

$60.4292° = \dfrac{\pi}{180} \times 60.4292 = \mathbf{1.0547\ radians}$

10.1.1 Types of angle

Some of the main types of angle are:

- acute angle: an angle less than 90° (Figure 10.1)
- right angle: an angle equal to 90° (Figure 10.2a)
- obtuse angle: an angle that is greater than 90° but less than 180° (Figure 10.3a)
- complementary angles: angles whose sum is 90°. For example, 40° and 50° are complementary angles as 40° + 50° = 90°.
- supplementary angles: angles whose sum is 180°. For example, 110° and 70° are supplementary angles as 110° + 70° = 180°.
- adjacent angles: angles on a straight line whose sum is 180°. In Figure 10.3b ∠A and ∠B are adjacent angles.

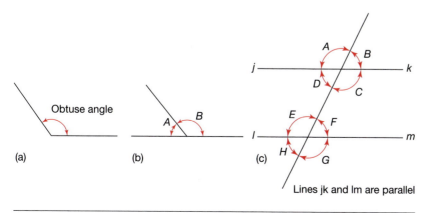

Lines jk and lm are parallel

Figure 10.3

- alternate angles: in Figure 10.3c, $\angle D$ and $\angle F$, and $\angle C$ and $\angle E$ are alternate angles. Alternate angles are equal: $\angle D = \angle F$ and $\angle C = \angle E$
- corresponding angles: in Figure 10.3c, $\angle B$ and $\angle F$; $\angle A$ and $\angle E$; $\angle C$ and $\angle G$; $\angle D$ and $\angle H$ are corresponding angles. Like alternate angles, corresponding angles are also equal. $\angle B = \angle F$, $\angle A = \angle E$, $\angle C = \angle G$, $\angle D = \angle H$.
- opposite angles: there are four pairs of opposite angles (also called vertically opposite angles) in Figure 10.3c. For example, $\angle A$ and $\angle C$ are opposite and equal: $\angle A = \angle C$, $\angle B = \angle D$, $\angle E = \angle G$, $\angle F = \angle H$.

Example 10.2

Calculate angles c, d and e shown in Figure 10.4. Line LM is parallel to line NP.

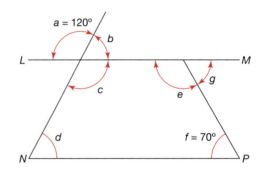

Figure 10.4

Solution:

LM is a straight line, therefore $\angle a + \angle b = 180°$

$\angle b = 180° - \angle a$

$\quad = 180° - 120° = 60°$

As $\angle b$ and $\angle d$ are corresponding angles, $\angle b = \angle d$

Therefore $\angle \boldsymbol{d} = \boldsymbol{60°}$

$\angle g = \angle f$, as these are alternate angles

$\angle g = 70°$

Now $\angle g + \angle e = 180°$

or $\angle e = 180° - \angle g$

Hence $\angle \boldsymbol{e} = 180° - 70° = \boldsymbol{110°}$

$\angle \boldsymbol{c} = \angle a = \boldsymbol{120°}$ (angles c and a are opposite and equal)

10.2 Polygons

Figures bounded by straight lines are called plane figures. Plane figures have only two dimensions, i.e. length and width. Shapes or figures made by straight lines are also called polygons, some of which are: triangles, rectangles, squares, trapeziums and pentagons.

10.3 Triangles

Triangles are plane figures bounded by three straight lines. A triangle is a very stable geometric shape and it is not possible to distort it without changing the length of one or more sides. Roof trusses and trussed rafters are used in the construction of roofs to dispose of rainwater and snow quickly. Their triangular shape also provides stability, which is one of the important requirements of a roof structure. Triangles also find use in the construction of multi-storey steel frames. Triangulation, a process used to convert the rectangular grids of steel frames into triangles, is necessary to make the frames and hence the buildings more stable.

10.3.1 Types of triangle (Δ)

The conventional method of denoting the angles of a triangle is to use capital letters, e.g. *A*, *B*, *C*, *P*, *R*, etc. The sides are given small letters; side *a* opposite angle *A*, side *b* opposite angle *B* and so on, as illustrated in Figure 10.5.

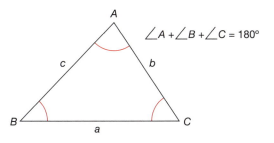

$$\angle A + \angle B + \angle C = 180°$$

Figure 10.5

Figure 10.5 shows a triangle which has each of its angles as less than 90°. This type of triangle is known as an *acute angled triangle*. Other types of triangles are described below:

- obtuse angled triangle: has one angle more than 90° (Figure 10.6a)
- right-angled triangle: one of the angles is equal to 90° (Figure 10.6b)
- equilateral triangle: has equal sides (Figure 10.6c). Because of the equal sides the angles are equal as well, each being 60°.
- isosceles triangle: has two sides and two equal angles (Figure 10.6d).
- scalene triangle: has all angles of different magnitude and all sides of different length. Acute angled, obtuse angled and right-angled triangles can be scalene triangles as well.

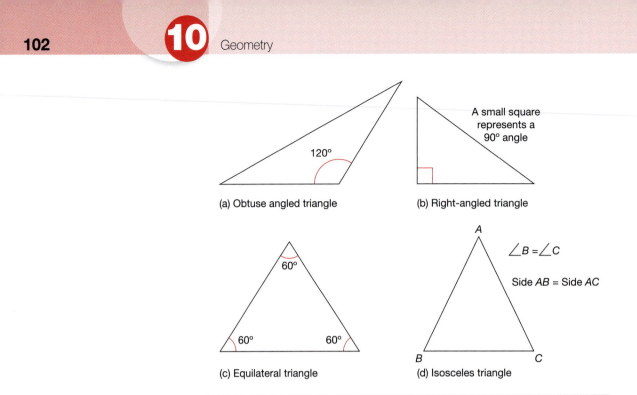

(a) Obtuse angled triangle

(b) Right-angled triangle

A small square represents a 90° angle

(c) Equilateral triangle

(d) Isosceles triangle

$\angle B = \angle C$

Side AB = Side AC

Figure 10.6

10.3.2 Theorem of Pythagoras

In a right-angled triangle (Figure 10.6b) the longest side, known as the hypotenuse, is opposite the right angle.

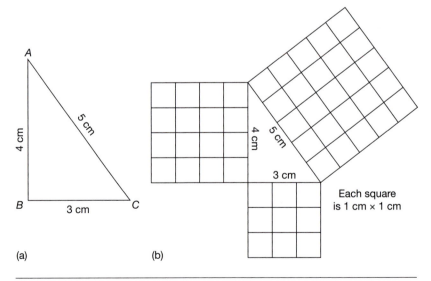

(a)   (b)

Figure 10.7

Figure 10.7a shows a right-angled triangle with sides 3 cm and 4 cm long. Side *AC*, the hypotenuse, is 5 cm long, which can be proved mathematically by drawing 1 cm × 1 cm squares on all sides of the triangle, as illustrated in Figure 10.7b.

Number of squares on side AB = 16

$$= 4^2 = (\text{side } AB)^2 \text{ or } (AB)^2$$

Number of squares on side BC = 9

$$= 3^2 = (BC)^2$$

Number of squares on side AC = 25 = 5^2 or $(AC)^2$

The area of each square is 1 cm², therefore the total area of the large square on any side is equal to the number of squares drawn on that side. As all squares are equal, we can say that:

Area of squares on side AC = Sum of area of squares on sides AB and BC

or $AC^2 = AB^2 + BC^2$ (1)

Take the square root of both sides

$$AC = \sqrt{(AB)^2 + (BC)^2}$$
$$= \sqrt{(4)^2 + (3)^2}$$
$$= \sqrt{16 + 9}$$
$$= \sqrt{25} = 5 \text{ cm}$$

Equation 1 can be used to determine any side of a right-angled triangle if the other two are known.

Example 10.3

Figure 10.8 shows three right-angled triangles. For each triangle, find the unknown side.

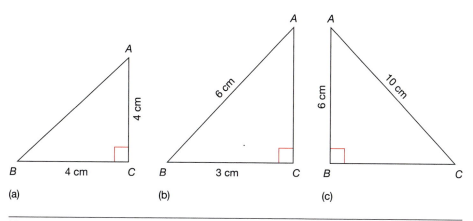

(a)   (b)   (c)

Figure 10.8

Solution:

Use Pythagoras' Theorem for each triangle.

(a) In $\triangle ABC$, $AC = BC = 4$ cm. AB is the hypotenuse:

$$AB^2 = BC^2 + AC^2$$
$$= 4^2 + 4^2$$
$$= 16 + 16 = 32$$

Taking the square root of both sides

$$AB = \sqrt{32} = \mathbf{5.657\ cm}$$

(b) In $\triangle ABC$, $AB = 6$ cm and $BC = 3$ cm. AB is the hypotenuse:

$$AB^2 = BC^2 + AC^2$$

Side AC is to be determined; transpose the above equation to make AC^2 the subject:

$$AB^2 - BC^2 = AC^2 \text{ or } AC^2 = AB^2 - BC^2$$
$$AC^2 = 6^2 - 3^2$$
$$= 36 - 9$$
$$= 27$$
$$AC = \sqrt{27} = \mathbf{5.196\ cm}$$

(c) Given: $AB = 6$ cm, $AC = 10$ cm. AC is the hypotenuse:

$$AC^2 = AB^2 + BC^2$$

Transpose the above equation to make BC the subject

$$AC^2 - AB^2 = BC^2$$
$$\text{or } BC^2 = AC^2 - AB^2$$
$$= 10^2 - 6^2$$
$$= 64$$
$$BC = \sqrt{64} = \mathbf{8\ cm}$$

10.3.3 Similar triangles

Two triangles are similar if the angles of one triangle are equal to the angles of the other triangle. Similar triangles are different in size but have the same shape (see Figure 10.9) and their sides are proportional.

$$\frac{AB}{DE} = \frac{BC}{EF} = \frac{CA}{FD}$$

In similar triangles, one of the following conditions will apply:

- Two angles in one triangle are equal to two angles in the other triangle.
- Two sides of one triangle are proportional to two sides of the other triangle and the angles included between the two sides are equal.
- The three sides of one triangle are proportional to the three sides of the other triangle.

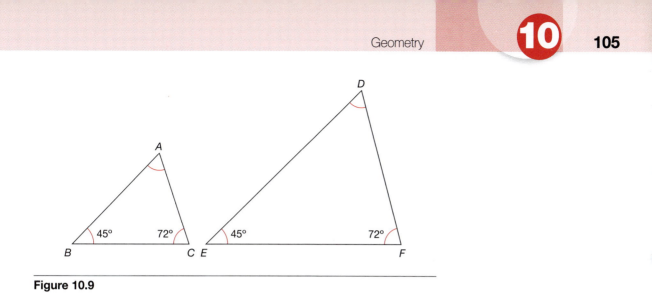

Figure 10.9

Example 10.4

Figure 10.10 shows △*ADE* with a line *BC* forming another triangle. If line *BC* is parallel to line *DE* and the sides of the triangle are as shown, show that:

(a) Triangles *ABC* and *ADE* are similar.

(b) Calculate *AD* and *DE*.

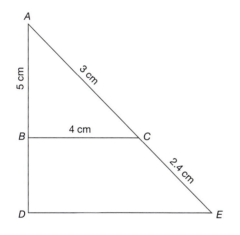

Figure 10.10

Solution:

(a) Line *BC* is parallel to line *DE*, therefore:

∠*ABC* = ∠*BDE*, and ∠*ACB* = ∠*CED*

∠*A* is common to both triangles.

Hence △*ABC* and △*ADE* are similar (three angles equal)

(b) As △*ABC* and △*ADE* are similar, their sides should be proportional:

$$\frac{BC}{DE} = \frac{AC}{AE}$$

$$\frac{4}{DE} = \frac{3}{5.4} \quad (AE = 3 + 2.4 = 5.4)$$

Transposing, $3DE = 4 \times 5.4$

$$\therefore DE = \frac{4 \times 5.4}{3} = \textbf{7.2 cm}$$

$$\frac{AB}{AD} = \frac{AC}{AE}$$

$$\frac{5}{AD} = \frac{3}{5.4}$$

Transposing, $3AD = 5 \times 5.4$

$$\text{or } AD = \frac{5 \times 5.4}{3} = \textbf{9.0 cm}$$

Example 10.5

Figure 10.11 shows a Fink roof truss. Find the length of members AB and BE.

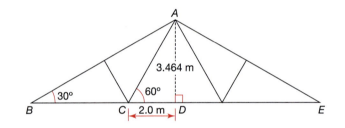

Figure 10.11

Solution:

From point A draw a vertical line to meet *BE* at point *D*. As the truss is symmetrical, $BE = 2BD$. We will calculate *BD* and multiply it by 2 to get *BE*.

Consider $\triangle ABD$ and $\triangle ACD$. Angle *ADB* is a right angle and common to both triangles.

$$\angle BAD = 180° - 90° - 30° = 60°$$

$$\angle CAD = 180° - 90° - 60° = 30°$$

$\triangle ACD$ is a right-angled triangle

$$AC^2 = AD^2 + CD^2$$

$$= 3.464^2 + 2^2$$

$$= 16$$

$$AC = \sqrt{16} = 4 \text{ m}$$

$\triangle ABD$ and $\triangle ACD$ are similar as the three angles are equal (Figure 10.12). Therefore, their sides are proportional.

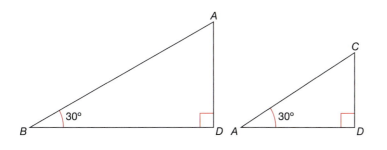

Figure 10.12

$$\frac{AD}{CD} = \frac{BD}{AD}$$

$$\frac{3.464}{2} = \frac{BD}{3.464}$$

$2BD = 3.464 \times 3.464$

$BD = (3.464 \times 3.464) \div 2 = 6.0$ m

$BE = 2 \times BD = 2 \times 6.0 =$ **12 m**

Consider $\triangle ABD$ and $\triangle ACD$ again:

$$\frac{AB}{CA} = \frac{AD}{CD}$$

$$\frac{AB}{4} = \frac{3.464}{2}$$

$2AB = 3.464 \times 4$

$AB = (3.464 \times 4) \div 2 =$ **6.928 m**

10.4 Quadrilaterals

Figures bounded by four straight lines are called quadrilaterals. Rectangle, square, trapezium, parallelogram and kite are called special quadrilaterals (Figure 10.13). A line joining the opposite corners of a quadrilateral is called a diagonal. There are two diagonals in a quadrilateral, and depending on the shape of the quadrilateral they may or may not be equal. A diagonal divides a quadrilateral into two triangles, so the sum of the internal angles of a quadrilateral is 360°.

In a rectangle each angle is equal to 90° and the opposite sides are equal (Figure 10.13a).

Side AB = side DC, and side DA = side CB.

As explained earlier, lines DB and CA are called diagonals, which can be calculated by using Pythagoras' Theorem:

$CA^2 = AB^2 + BC^2$ ($\triangle DAB$ and $\triangle CAB$ are right-angled triangles)

$DB^2 = DA^2 + AB^2$

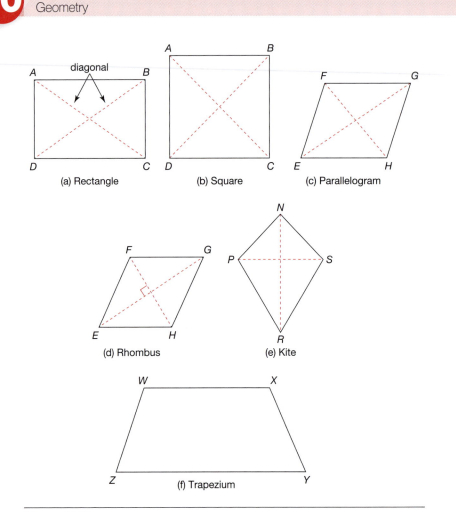

Figure 10.13

A square (Figure 10.13b) is a special form of rectangle in which all four sides are equal. Other properties are similar to those of a rectangle.

A quadrilateral with its opposite sides parallel and equal is called a parallelogram (Figure 10.13c). The sum of all internal angles is 360°, but only the opposite angles are equal. The rectangle is a special form of parallelogram with all internal angles equal. The diagonals in a parallelogram are not equal, and unlike the rectangle cannot be determined using Pythagoras' Theorem.

Side *EH* is parallel and equal to side *FG*.

Side *EF* is parallel and equal to side *HG*.

A rhombus is basically a parallelogram in which all four sides are equal (Figure 10.13d). The diagonals are not equal, but they bisect each other at right angles.

A kite is a quadrilateral that has two pairs of equal sides. The equal sides are adjacent to each other, as shown in Figure 10.13e.

Side *NP* = side *NS*

Side *RP* = side *RS*

A trapezium (Figure 10.13f) is a quadrilateral with a pair of parallel sides which are not of equal length:

Sides *WX* and *ZY* are parallel but unequal

Example 10.6

The size of a rectangle is 12 cm × 9 cm. Calculate the length of each diagonal.

Solution:

Figure 10.14 shows the dimensions of the rectangle.

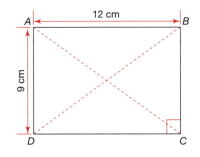

Figure 10.14

Use Pythagoras' Theorem to determine the length of one diagonal.

$$AC = BD = \sqrt{12^2 + 9^2}$$
$$= \sqrt{225} = \textbf{15 cm}$$

Example 10.7

Figure 10.15 shows a parallelogram *EFGH*. Calculate angles *a*, *b*, *c* and *d*.

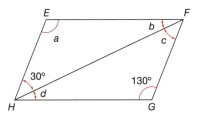

Figure 10.15

Solution:

In a parallelogram the opposite angles are equal and the sum of the internal angles is equal to 360°.

$$\angle a = \angle FGH = \textbf{130°}$$
$$\angle EFG = \angle EHG = \frac{360° - 130° - 130°}{2}$$
$$= 50°$$

$\angle d = \angle b = 50° - 30° = \mathbf{20°}$

In $\triangle HFG$, $\angle c = 180° - 130° - 20° = \mathbf{30°}$

10.5 Sum of the angles in a polygon

The sum of the angles in a polygon can be determined using a number of techniques, but in this section only two of them are described.

The first method involves the splitting of the given polygon into triangles. The sum of the angles will be equal to the number of triangles formed multiplied by 180°. This is shown in Figure 10.16.

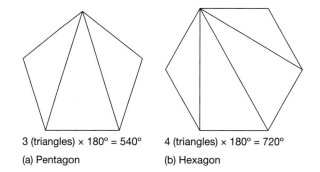

3 (triangles) × 180° = 540°   4 (triangles) × 180° = 720°

(a) Pentagon   (b) Hexagon

Figure 10.16

The second method involves the use of the formula:

The sum of the internal angles of a polygon = $(2n - 4)$ right angles where n is the number of the sides of the polygon

Table 10.1 shows some of the calculations

Table 10.1

Polygon	Number of sides	Number of right angles ($2n - 4$)	Sum of the internal angles
Triangle	3	2 × 3 − 4 = 2	2 × 90 = 180°
Quadrilaterals (rectangle, square, etc.)	4	2 × 4 − 4 = 4	4 × 90 = 360°
Pentagon	5	2 × 5 − 4 = 6	6 × 90 = 540°
Hexagon	6	2 × 6 − 4 = 8	8 × 90 = 720°
Heptagon	7	2 × 7 − 4 = 10	10 × 90 = 900°
Octagon	8	2 × 8 − 4 = 12	12 × 90 = 1080°
Nonagon	9	2 × 9 − 4 = 14	14 × 90 = 1260°
Decagon	10	2 × 10 − 4 = 16	16 × 90 = 1440°

10.6 The circle

If we take a compass, open it by a certain amount (*OA*) and draw a curved line, the result is a shape as shown in Figure 10.17a. The plane figure enclosed by the curved line is called a circle. Line *OA* is called the radius (*r*) of the circle. Every point on the curved line is equidistant from the centre of the circle. In Figure 10.17b, *OA*, *OB* and *OC* are the radii (plural of radius) of the circle and are equal.

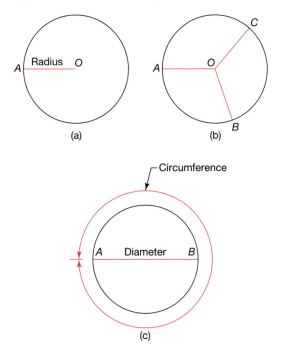

Figure 10.17

The length of the curved line (Figure 10.17c) is called the circumference (*c*). A straight line that passes through the centre with its ends touching the circumference is called a diameter (*d*), e.g. line AB in Figure 10.17c is a diameter.

$d = 2 \times \text{radius}$
or $d = 2r$

A diameter divides the circle into two equal parts, each known as a semicircle. The ratio of the circumference to the diameter of a circle is a constant and known by the Greek letter π (pi):

$$\frac{\text{Circumference}}{\text{Diameter}} = \pi, \text{ or } \frac{c}{d} = \pi$$

$$\therefore c = \pi d$$

Also:

$$c = \pi \times 2r = 2\pi r \qquad (d = 2r)$$
$$\pi = 3.14159265 \qquad \text{(to 8 d.p.)}$$
$$= 3.142 \qquad \text{(to 3 d.p.)}$$

Any straight line from circumference to circumference, such as line EF in Figure 10.18a, is called a chord. A chord divides a circle into two segments. The smaller segment is called the minor segment and the larger one is called the major segment.

The plane figure contained by the two radii and an arc of the circumference is called a sector. If the angle contained by the two radii is less than 180°, the sector is called the minor sector, as shown in Figure 10.18b.

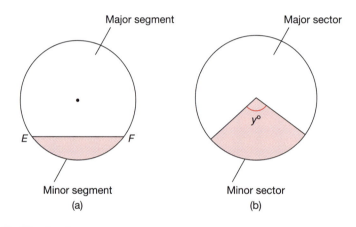

Figure 10.18

Area of minor sector = area of the circle $\times \dfrac{y°}{360°}$

Example 10.8

Calculate the circumference of a circle if its radius is 20 cm.

Solution:

Radius r = 20 cm

The circumference of the given circle = $2\pi r$
$$= 2\pi \times 20 = \mathbf{125.66\ cm}$$

Exercise 10.1

The solutions to Exercise 10.1 can be found in Appendix 2.

1. Convert: (a) 72°38′51″ into degrees (decimal measure)
   (b) 63.88° into degrees, minutes and seconds.
   (c) 30°48′50″ into radians.

2. Calculate angles x, y and z shown in Figure 10.19. Line *AB* is parallel to line *CD*.

3. For each of the right-angled triangles shown in Figure 10.20, find the unknown side.

Figure 10.19

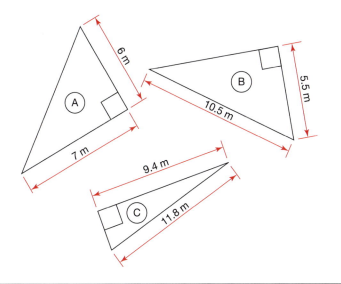

Figure 10.20

4. In Figure 10.21, line *XY* is parallel to line *PN* and creates another triangle, i.e. Δ ***MRQ***:

    (a) Show that triangles *MNP* and *MRQ* are similar.

    (b) Calculate the lengths of *PQ* and *RN*.

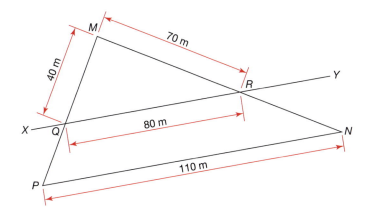

Figure 10.21

5. Find the lengths of all the members of the King Post roof truss shown in Figure 10.22.

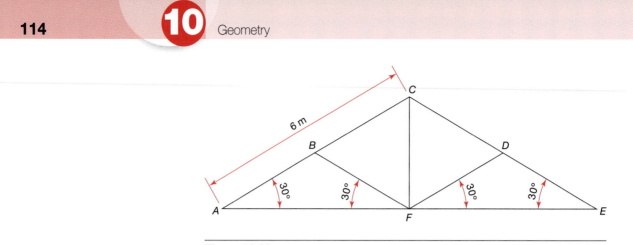

Figure 10.22

6. A building site has two roads at right-angles to each other, each forming a boundary to the site. If the boundaries are 26 m and 38 m long, and the site is triangular, what is the length of the third boundary?

7. A parallelogram is shown in Figure 10.23. Find the angles *A*, *B* and *C*.

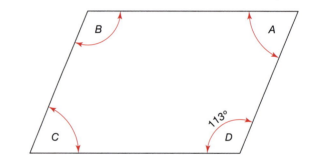

Figure 10.23

8. Calculate the circumference of a circle having a radius of 5.6 m.

Answers to Exercise 10.1

1. (a) 72.6475°; (b) 63°52′48″; (c) 0.5378 radians
2. $\angle x = 104°$, $\angle y = 107°$, $\angle z = 73°$
3. $A = 9.220$ m, $B = 8.944$ m, $C = 7.133$ m
4. (a) $\angle M$ is common to both $\triangle$s; $\angle Q = \angle P$; $\angle R = \angle N$; $\therefore \triangle$s are similar.
   (b) $PQ = 15$ m and $RN = 26.25$ m
5. $AB = BC = CD = DE = DF = BF = CF = 3.0$ m; $AF = FE = 5.196$ m
6. 46.043 m
7. $\angle B = 113°$, $\angle A = 67°$, $\angle C = 67°$
8. 35.186 m

CHAPTER 11

Areas (1)

Learning outcomes:

(a) Calculate the areas of triangles, quadrilaterals and circles

(b) Identify and use the correct units

(c) Solve practical problems involving area calculation

11.1 Introduction

Area is defined as the amount of space taken up by a two-dimensional figure. The geometrical properties of triangles, quadrilaterals and circles have been explained in Chapter 10. A summary of the formulae used in calculating the areas and other properties of these geometrical shapes is given in Table 11.1. The units of area used in metric systems are: mm^2, cm^2, m^2 and km^2.

Table 11.1

Shape	Area and other properties
Triangle 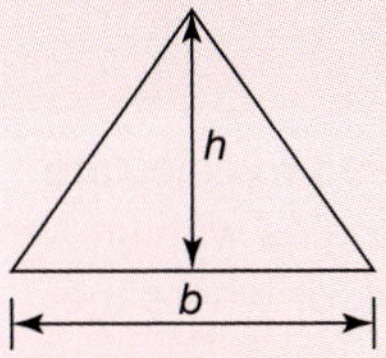	$Area = \dfrac{b \times h}{2}$
Rectangle 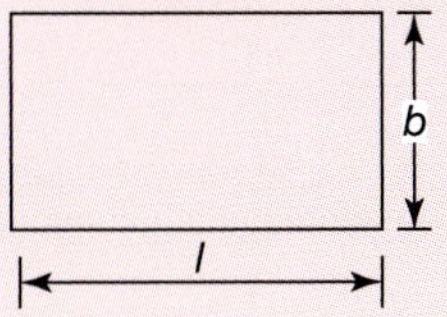	$Area = l \times b$ $Perimeter = 2l + 2b = 2(l + b)$

(Continued)

Table 11.1 *(Continued)*

Shape	Area and other properties
Square	Area = $l \times l = l^2$ Perimeter = $4l$
Trapezium	Area = $\dfrac{1}{2}(a + b) \times h$
Parallelogram	Area = $l \times h$
	Area = πr^2 Circumference = $2\pi r$

11.2 Area of triangles

There are many techniques and formulae that can be used to calculate the area of triangles. In this section we consider the triangles with known measurements of the base and the perpendicular height, or where the height can be calculated easily.

Example 11.1

Find the area of the triangles shown in Figure 11.1.

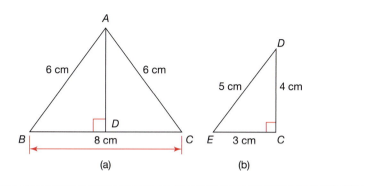

Figure 11.1

Solution:

(a)

Area of triangle $ABC = \dfrac{\text{base} \times \text{height}}{2}$

Base $BC = 8$ cm

We need to calculate height AD, which has not been given. As sides AB and AC are equal, BD must be equal to DC. Therefore, $BD = DC = 4$ cm. Now we can use Pythagoras' Theorem to calculate height AD:

$$(AD)^2 = (AB)^2 - (BD)^2$$
$$= 6^2 - 4^2$$
$$= 36 - 16 = 20$$

Therefore

$$AD = \sqrt{20} = 4.47 \text{ cm}$$

Area of triangle $ABC = \dfrac{8 \times 4.47}{2} = \mathbf{17.88 \text{ cm}^2}$

(b)

Area of triangle $DEF = \dfrac{\text{base} \times \text{height}}{2}$

$$= \dfrac{3 \times 4}{2} = \mathbf{6 \text{ cm}^2}$$

11.3 Area of quadrilaterals

A plane figure bounded by four straight lines is called a quadrilateral. The calculation of area of some of the quadrilaterals is explained in this section.

Example 11.2

Find the area of the shapes shown in Figure 11.2.

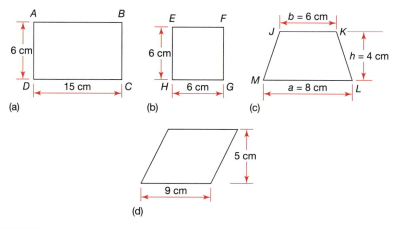

(a) (b) (c)

(d)

Figure 11.2

Solution:

(a)

Area of a rectangle = length × width

Length = 15 cm, and width = 6 cm

Area of rectangle *ABCD* = 15 × 6 = **90 cm²**

(b)

In a square, the length is equal to the width. Therefore:

Area of a square = length × length (or length²)

Area of square *EFGH* = (6)² = **36 cm²**

(c)

Area of trapezium *JKLM* = $\frac{1}{2}(a + b) \times h$

$$= \frac{1}{2}(8 + 6) \times 4$$

$$= \frac{1}{2} \times 14 \times 4 = \textbf{28 cm}^2$$

(d)

Area of a parallelogram = length × height

Length = 9 cm, and height = 5 cm

Area of the parallelogram = 9 × 5 = **45 cm²**

11.4 Area of circles

Some of the important properties of circles are explained in Chapter 10. The following formula is used to calculate the area of a circle:

Area of a circle = πr^2, where *r* is the radius of the circle.

Example 11.3

Figure 11.3 shows a circle having a diameter of 4 m. Find:

(a) the area of the circle
(b) the area of the major and minor sectors.

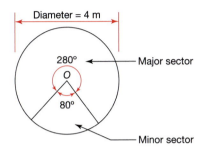

Figure 11.3

Solution:

Radius (r) of the circle $= \dfrac{\text{Diameter}}{2} = \dfrac{4}{2} = 2$ m

(a)

Area of the circle $= \pi r^2 = \pi \times 2^2 = $ **12.57 m²**

(b)

Area of the major sector $= \pi r^2 \times \dfrac{280°}{360°}$

$= \pi \times 2^2 \times \dfrac{280}{360} = $ **9.774 m²**

Area of the minor sector $= \pi r^2 \times \dfrac{80°}{360°}$

$= \pi \times 2^2 \times \dfrac{80}{360} = $ **2.793 m²**

Example 11.4

Figure 11.4 shows a circular concrete path (shaded) around a lawn.
Find the area of the path.

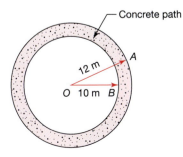

Figure 11.4

Solution:

Area of the path = area of the bigger circle – area of the smaller circle

$$= \pi \times (12)^2 - \pi \times (10)^2$$

$$= 452.39 - 314.16 = \textbf{138.23 m}^2$$

11.5 Application of area to practical problems

The concepts learned from Sections 11.2–11.4 may be used to solve many practical problems, some of which are dealt with in this section.

Example 11.5

Figure 11.5 shows the cross-section of a steel girder. Calculate its area in cm^2 and mm^2.

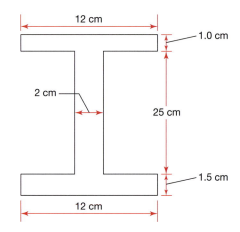

Figure 11.5

Solution:

The steel section shown in Figure 11.5 can be divided into three parts as shown in Figure 11.6

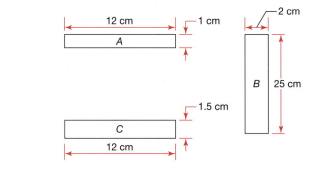

Figure 11.6

Area of the steel section = Area A + Area B + Area C

$\qquad\qquad$ = 12 × 1 + 25 × 2 + 12 × 1.5

$\qquad\qquad$ = 12 + 50 + 18

$\qquad\qquad$ = **80 cm²**

$\qquad\qquad$ = 80 × 100 = **8000 mm²**   (1 cm² = 100 mm²)

(Note: where the cross-section of a component is given, the area is also called the cross-sectional area.)

Example 11.6

The floor of a kitchen, measuring 3.20 m × 3.00 m, is to be finished with marble-effect floor tiles, each measuring 330 mm × 330 mm. If there are nine tiles in a pack, calculate the number of packs required. Allow 10% for wastage.

Solution:

Area of the floor = 3.20 × 3.00 = 9.60 m²

Area of one tile $\ $ = 330 mm × 330 mm

$\qquad\qquad$ = 0.33 m × 0.33 m = 0.1089 m²

As there are nine tiles in a pack, each pack will cover 0.1089 × 9 or 0.98 m² of the floor.

Number of packs required = $\dfrac{\text{Area of the floor}}{\text{Area of tiles in one pack}}$

$\qquad\qquad = \dfrac{9.6}{0.98} = 9.8$

Wastage = 10% of 9.8 = 9.8 × $\dfrac{10}{100}$ = 0.98

Number of packs required (including 10% for wastage) = 9.8 + 0.98

$\qquad\qquad\qquad\qquad\qquad\qquad\qquad\qquad\qquad$ = **10.78, so 11.**

Example 11.7

The floor of a room measuring 6.0 m × 4.2 m is to be covered with laminate flooring. If one pack covers an area of 2.106 m², calculate the number of packs required. Consider wastage to be 5%.

Solution:

Area of the floor = 6.0 m × 4.2 m

$\qquad\qquad$ = 25.20 m²

Number of packs required = $\dfrac{\text{Floor area}}{\text{Area covered by one pack}}$

$\qquad\qquad = \dfrac{25.2}{2.106} = 11.97$

Wastage of 5% = $\dfrac{5}{100}$ × 11.97 = 0.6

Number of packs required = 11.97 + 0.6

= **12.57, so 13**

Example 11.8

The 3D view of a room is shown in Figure 11.7. The internal wall and ceiling surfaces require a coat of emulsion paint. Calculate their areas in m². Given: size of the door = 2 m × 1 m wide; height of the skirting board = 100 mm

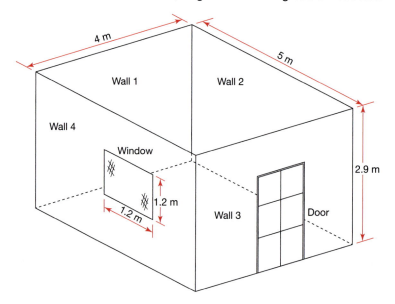

Figure 11.7

Solution:

The area of each wall will be calculated separately and added later to find the total wall area. The areas of the window, door and skirting board have to be excluded.

Area of wall 1 (see Figure 11.8a) = length × height

= 4.0 × (2.9 − 0.1)

= 4.0 × 2.8 = 11.2 m²

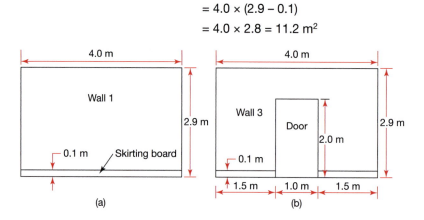

Figure 11.8

Area of wall 2 = length × height

$$= 5.0 \times (2.9 - 0.1)$$

$$= 5.0 \times 2.8 = 14.0 \text{ m}^2$$

Area of wall 3 = Gross area of the wall – Area of the door – Area of the skirting board

$$= 4.0 \times 2.9 - 2.0 \times 1.0 - (1.5 \times 0.1 + 1.5 \times 0.1)$$
(see Figure 11.8b)

$$= 11.6 - 2.0 - 0.3 = 9.3 \text{ m}^2$$

The details of wall 4 are shown in Figure 11.9.

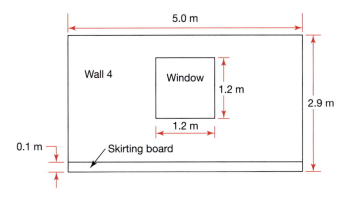

Figure 11.9

Area of wall 4 = gross area of the wall – area of the window – area of the skirting board

$$= 5.0 \times (2.9 - 0.1) - 1.2 \times 1.2$$

$$= 5.0 \times (2.8) - 1.44$$

$$= 12.56 \text{ m}^2$$

Total wall area = 11.2 + 14.0 + 9.3 + 12.56

$$= \textbf{47.06 m}^2$$

Area of the ceiling = 5.0 × 4.0 = **20 m²**

11.5.1 Cavity walls

The cavity wall of a domestic building consists of brickwork as outer leaf and blockwork as inner leaf, with the cavity between the two filled with Rockwool. The outer leaf of brickwork is 103 mm thick (also called half-brick-thick wall). The inner leaf of concrete blocks is 100 mm thick. The measurements of bricks and concrete blocks used in the area calculations are shown in Figure 11.10. The number of bricks and concrete blocks per m² are 60 and 10, respectively.

Example 11.9

A cavity wall 5.4 m long and 2.7 m high (Figure 11.10) is to be constructed using facing bricks and aerated concrete blocks. Find the number of bricks and blocks required. Allow 5% for wastage.

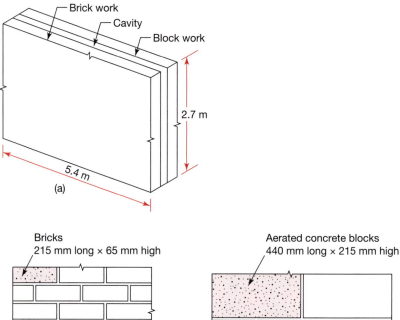

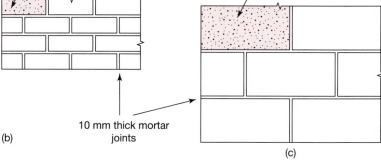

Figure 11.10

Solution:

Size of a brick without mortar joints = 215 × 102.5 × 65 mm

Size of a brick with mortar joints = 225 × 102.5 × 75 mm

Size of a block without mortar joints = 440 × 215 ×100 mm

Size of a block with mortar joints = 450 × 225 × 100 mm

Area of a brick = 225 × 75 = 16 875 mm²

Area of a block = 450 × 225 = 101 250 mm²

We must have compatible units of all components. Therefore, change the length and the height of the wall into millimetres.

Length of the wall = 5.4 m = 5400 mm

Height of the wall = 2.7 m = 2700 mm

Area of the wall = 5400 × 2700 = 14 580 000 mm²

Number of bricks required = $\dfrac{\text{Area of the wall}}{\text{Area of one brick}}$

$$= \dfrac{14\,580\,000}{16\,875} = 864$$

Wastage of 5% = $864 \times \dfrac{5}{100} = 43.2$, say 44

Total number of bricks = 864 + 44 = **908**

Number of blocks required = $\dfrac{14\,580\,000}{101\,250} = 144$

Wastage of 5% = $144 \times \dfrac{5}{100} = 7.2$, say 8

Total number of blocks = 144 + 8 = **152**

Exercise 11.1

The solutions to Exercise 11.1 can be found in Appendix 2.

1. Find the area of the shapes shown in Figure 11.11.

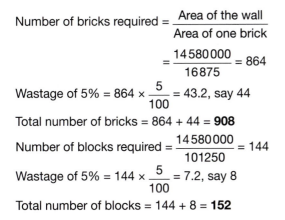

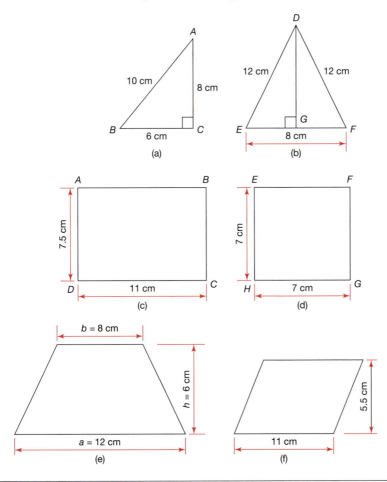

Figure 11.11

Diameter = 3.5 m

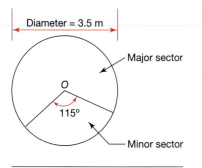
Major sector

O

115°

Minor sector

Figure 11.12

2. Figure 11.12 shows a circle having a diameter of 3.5 m. Find:
   (a) the area of the circle
   (b) the area of the major and minor sectors.

3. Figure 11.13 shows a circular concrete path (shaded) around a lawn. Find the area of the path.

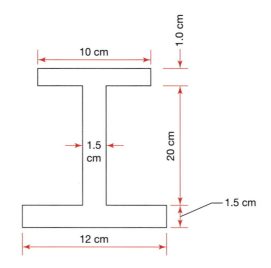
Concrete path

16 m

15 m

Figure 11.13

4. A concrete path is to be provided around a rectangular garden measuring 15 m in length and 11 m in width. Find the area of the path if it is 1.2 m wide.

5. Figure 11.14 shows the cross-section of a steel girder. Calculate its area in cm^2 and mm^2.

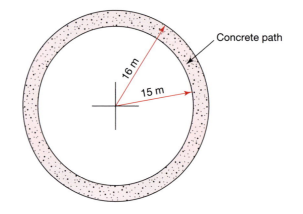
10 cm

1.0 cm

1.5 cm

20 cm

1.5 cm

12 cm

Figure 11.14

6. The floor of a kitchen, measuring 3.50 m × 3.00 m, is to be finished with marble effect floor tiles each measuring 330 mm × 330 mm. If there are nine tiles in a pack, calculate the number of packs required. Allow 10% for wastage.

7. Find the number of 200 × 100 × 50 mm concrete blocks required for paving the drive shown in Figure 11.15. Allow 10% for cutting and wastage.

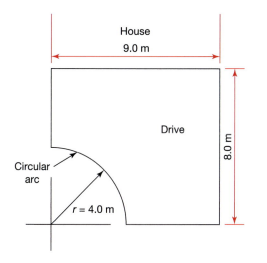

House
9.0 m

Drive

8.0 m

Circular arc

r = 4.0 m

Figure 11.15

8. The floor of a room, measuring 5.0 m × 4.0 m is to be covered with laminate flooring. If one pack covers an area of 2.106 m², calculate the number of packs required. Add 10% for wastage.

9. The 3D view of a room is shown in Figure 11.16. The internal wall and ceiling surfaces require a coat of emulsion paint. Calculate their areas in m². Given: size of the door = 2 m × 1 m wide; height of the skirting board = 100 mm.

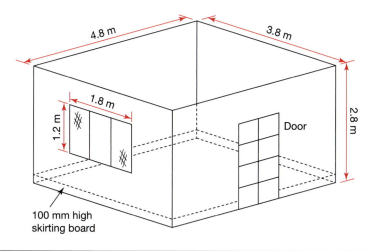

4.8 m   3.8 m

1.8 m   Door

1.2 m   2.8 m

100 mm high skirting board

Figure 11.16

10. Calculate the number of rolls of wallpaper required for the room shown in Figure 11.16. Each roll is 52 cm wide and 10.0 m long. Allow an extra 15% for wastage.

11. A cavity wall 5.0 m long and 2.4 m high is to be constructed using facing bricks and aerated concrete blocks. Find the number of bricks and blocks required. Allow 5% extra for wastage.

12. Figure 11.17 shows the gable end of a building. Calculate the number of bricks and aerated concrete blocks required for its construction. Allow an extra 5% for wastage.

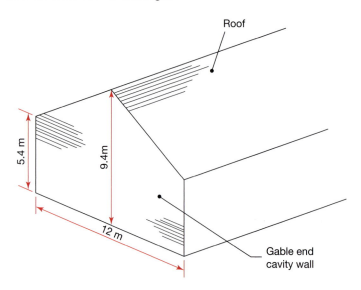

Figure 11.17

Answers to Exercise 11.1

1. (a) 24 cm²; (b) height = 11.31 cm; area = 45.255 cm²; (c) 82.5 cm²; (d) 49 cm²; (e) 60 cm²; (f) 60.5 cm²

2. Area (circle) = 9.621 m²; area (major sector) = 6.548 m²; area (minor sector) = 3.073 m²

3. 97.389 m²

4. 68.16 m²

5. 58 cm² or 5800 mm²

6. 12 packs

7. 3269 blocks

8. 11 packs

9. Wall area = 42.38 m², ceiling area = 18.24 m²

10. 10 rolls

11. 747 bricks, 125 blocks

12. 5526 bricks, 921 blocks

CHAPTER **12**

Volumes (1)

Learning outcomes:

(a) Identify cubes, cuboids, rectangular and triangular prisms, pyramids and cones

(b) Perform calculations to calculate the volumes of the above solids

(c) Perform calculations to determine the volume of concrete required for constructing different elements of a building

(d) Calculate the volume of cement, sand and gravel required to prepare a concrete mix

12.1 Introduction

Volume may be defined as the space in a three-dimensional object. This is different from the area, which is applicable to two-dimensional shapes. The most basic shapes in volume calculations are the cube and the cuboid. A cube is a three-dimensional figure which has six square faces. This means that the length, width and height are equal in a cube. In a cuboid, at least one of the sides will be different from the others. A typical example of a cuboid is an aerated concrete block.

To understand the concept of volume, divide a cuboid into smaller units, as shown in Figure 12.1. Each small unit, a cube, measures 1 cm × 1 cm × 1 cm.

Volume of a cuboid = length × width × height

or Volume, $V = l \times w \times h$

Volume of the unit cube = 1 cm × 1 cm × 1 cm

$$= 1 \text{ cm}^3$$

In Figure 12.1a there are 18 unit cubes. Therefore:

Volume of the cuboid = 1 cm³ × 18 = 18 cm³

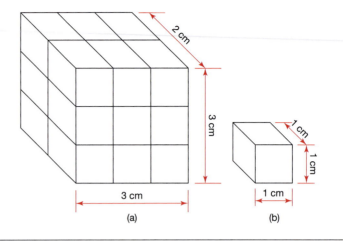

Figure 12.1

Using the formula:

$V = l \times w \times h$

$= 3 \times 2 \times 3 = 18 \text{ cm}^3$

Two other formulae that may also be used to calculate the volume of an object are:

Volume = area × length/width

Volume = area × height

These relationships are used to find the volume of triangular prisms, cylinders and other solids which have a uniform cross-section.

12.2 Volume of prisms, cylinders, pyramids and cones

The volume of a prism can be determined by using the formula explained in Section 12.1. Figure 12.2a and 12.2b show a rectangular and a triangular prism, respectively.

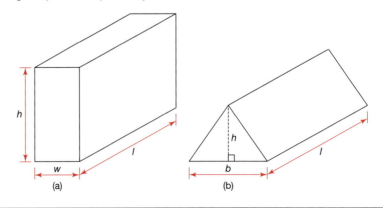

Figure 12.2

Volume of a rectangular prism = cross-sectional area × length

$$= (w \times h) \times l = lwh$$

Volume of a triangular prism = cross-sectional area × length

$$= \frac{b \times h}{2} \times l$$

The plan view of a cylinder (Figure 12.3) is a circle; its area is given by the formula, $A = \pi r^2$.

Volume of a cylinder = cross-sectional area × height

$$= \pi r^2 \times h = \pi r^2 h$$

A three-dimensional figure whose lateral sides are triangles is called a pyramid. The shape of a pyramid depends on the shape of its base (Figure 12.4). The base could be a rectangle, square, triangle or any other polygon, but in each case the top is a vertex. A regular pyramid has all its lateral sides equal.

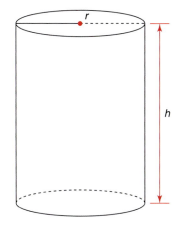

Figure 12.3

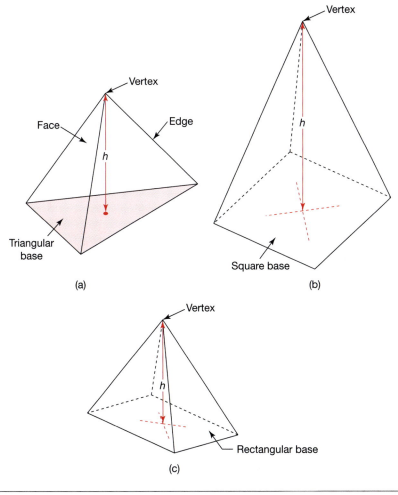

Figure 12.4

The calculation of the volume of a pyramid can be generalised as:

Volume of a pyramid = $\dfrac{1}{3}$ × area of the base × perpendicular height

Volume of a pyramid with rectangular base = $\dfrac{1}{3}(l \times w)h$

A cone and a pyramid are similar except that the base of a cone is a circle. Depending on the geometry of a cone it can be a right cone or an oblique cone:

right cone: the altitude intersects the base of the cone at its centre;

oblique cone: the altitude does not intersect the base of the cone at its centre.

Volume of a cone = $\dfrac{1}{3}$ × area of the base × perpendicular height

$$= \dfrac{1}{3}\pi r^2 h$$

Example 12.1

Find the volume of the solids shown in Figure 12.5.

Solution:

(a) Length = 15 cm; width = 10 cm; height = 20 cm

Volume of the cuboid = 15 × 10 × 20 = **3000 cm³**

(b) Diameter of the cylinder = 200 mm

Radius $r = \dfrac{200}{2} = 100$ mm

Volume of the cylinder = $\pi r^2 \times h$

$$= \pi (100)^2 \times 200$$

$$= \textbf{6 283 200 mm}^3 \text{ or } \textbf{6.283} \times \textbf{10}^6 \textbf{ mm}^3$$

(c) Volume of triangular prism = cross-sectional area × length

$$= \dfrac{80 \times 60}{2} \times 100$$

$$= \textbf{240 000 mm}^3$$

(d) The base of the pyramid is a square (0.3 × 0.3 m)

Volume of the pyramid = $\dfrac{1}{3}$ × area of the base × perpendicular height

$$= \dfrac{1}{3} \times (0.3 \times 0.3) \times 0.4$$

$$= \textbf{0.012 m}^3$$

(e) Diameter of the base = 12 cm

Radius of the base $= \dfrac{12}{2} = 6$ cm

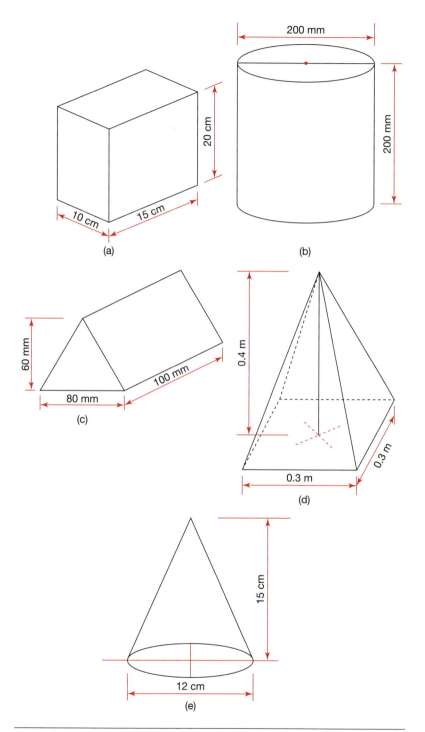

Figure 12.5

Volume of the cone $= \dfrac{1}{3} \times$ area of the base $\times$ perpendicular height

$$= \dfrac{1}{3} \times \pi(6)^2 \times 15 = \textbf{565.5 cm}^3$$

Example 12.2

Figure 12.6 shows the measurements of the roof of a building. Calculate the volume of the loft space.

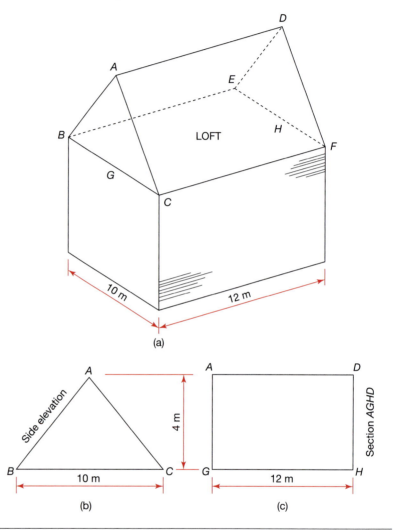

Figure 12.6

Solution:

The shape of the roof is similar to a triangular prism. The cross-section of the roof is a triangle with a base of 10 m and perpendicular height of 4 m.

Volume of the loft space = area of △ABC × length

$$= \frac{10 \times 4}{2} \times 12 = \textbf{240 m}^3$$

Example 12.3

The dimensions of a concrete pipe are shown in Figure 12.7. Find the volume of concrete used in its manufacture.

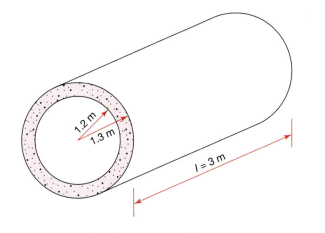

Figure 12.7

Solution:

The cross-section of the pipe is shown in Figure 12.7. To find the volume of concrete used, calculate the cross-sectional area of the concrete and multiply this by the length.

Cross-sectional area of concrete pipe $= \pi R^2 - \pi r^2$

$$= \pi (1.3)^2 - \pi (1.2)^2$$
$$= 5.31 - 4.52 = 0.79 \text{ m}^3$$

Volume of concrete = cross-sectional area × length

$$= 0.79 \times 3.0 \text{ m} = \textbf{2.37 m}^3$$

Example 12.4

The roof of a building has been designed as a pyramid (Figure 12.8). Find the volume of the space enclosed by the building.

Solution:

The building can be divided into a cuboid and a pyramid.

Volume of the cuboid $= 40 \times 40 \times 5 = 8000 \text{ m}^3$

Volume of the pyramid $= \dfrac{1}{3} \times (40 \times 40) \times 7$

$$= 3733.33 \text{ m}^3$$

Total volume $= 8000 + 3733.33 = \textbf{11 733.33 m}^3$

Example 12.5

The cross-section of an embankment is trapezoidal (trapezium) in shape. The measurements at the top and the base of the

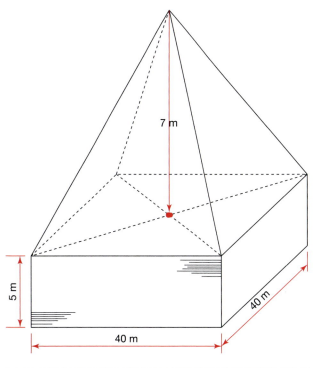

Figure 12.8

embankment are 10.0 m and 16.0 m, respectively. The height of the embankment is 2.0 m and the length 200 m. If the cross-section remains constant throughout the length of the embankment, find the volume of the soil:

(a) in its loose state (the soil bulks by 15%);

(b) in its compact state, after the construction of the embankment.

Solution:

Consider the embankment to be a prism as its cross-section is uniform throughout the length. Find the cross-sectional area of the embankment (trapezium) and multiply by the length to find the answer to part (b) of the question. Add 15% to find the volume of the earth in its loose state.

(a) Cross-sectional area of the embankment (trapezium) $= \dfrac{10 + 16}{2} \times 2.0$

$$= 13 \times 2.0 = 26 \text{ m}^2$$

Volume of soil in compact state $= 26 \times 200 = $ **5200 m³**

(b) Increase in the volume of soil due to bulking $= \dfrac{15}{100} \times 5200 = 780 \text{ m}^3$

Volume of soil in its loose state $= 5200 + 780 = $ **5980 m³**

12.3 Mass, volume and density

The density of building materials is an important property that a structural engineer has to consider for calculating the forces acting on a building. The self-weight of building components is permanent and known as the dead load. The live load, on the other hand, keeps changing, and typical examples are the weight of people using a building and wind force acting on a building.

Mass of a material/component = density × volume

Example 12.6

A water storage cistern measuring 2.5 m × 2.0 m × 1.2 m high is to be provided in a building.

(a) Calculate the volume of water when the cistern is filled to 15 cm from the top rim.

(b) Use the volume of water calculated in part (a) to determine the mass of water. The density of water is 1000 kg/m³.

(c) Find the load (force) on the joists supporting the cistern. 1 kg = 9.8 Newtons (N)

Solution:

(a) Convert 15 cm into metres. To do this divide 15 by 100 as

$$1 \text{ cm} = \frac{1}{100} \text{ m}$$

$$15 \text{ cm} = \frac{15}{100} \text{ m} = 0.15 \text{ m}$$

Depth of water in the cistern = 1.20 − 0.15 = 1.05 m

Volume of water = lwh

$$= 2.5 \times 2.0 \times 1.05 = \textbf{5.25 m}^3$$

(b) Density of water = 1000 kg/m^3. This means that 1 m^3 of water has a mass of 1000 kg.

Mass of 5.25 m^3 of water = 5.25 × 1000 = **5250 kg**

(c) Load (or force) on the joists = 5250 × 9.8 = **51 450 N**

12.4 Concrete mix and its constituents

Concrete is one of the important materials used in the construction of buildings and civil engineering structures. The constituents of a concrete mix are cement, fine aggregates (sand), coarse aggregates (gravel, blast-furnace slag, etc.) and water. Water is added for the chemical reaction between cement and water to take place and produce a solid concrete from a semi-fluid state. The amounts of these constituents may be determined by considering either their volume or mass. If the proportions of cement, sand and gravel are measured by volume, the concrete mix is known as a nominal mix. Mixing by volume does not take into account the moisture content of the aggregates, therefore, nominal mixes are used only for minor work. Typical examples of nominal mixes are 1:2:4 concrete and 1:3:6 concrete. The 1:2:4 concrete means that it is prepared by mixing one part cement, two parts fine aggregates and four parts coarse aggregates. The volume of water is about 50–60% of the volume of cement. In this section only nominal mixes are considered.

If we want to prepare 1 m^3 of 1:2:4 concrete mix, the quantities of dry materials may be calculated as shown below. The volume of cement in a 25 kg bag is 0.0166 m^3.

Assuming the volume of water to be 55% of the volume of cement, the proportions of the concrete mix may be written as 1:2:4:0.55. The total of these proportions is 7.55.

Volume of cement = $\frac{1}{7.55} \times 1 = 0.132$ m^3

Volume of fine aggregates = $\frac{1}{7.55} \times 2 = 0.265$ m^3

Volume of coarse aggregates = $\frac{1}{7.55} \times 4 = 0.53$ m^3

Volume of water = $\frac{1}{7.55} \times 0.55 = 0.073$ m^3 or 73 litres

The coarse aggregates contain a lot of air voids which are filled by the other ingredients when they are mixed. The final volume of the concrete mix will be less than 1 m^3. The volume of the dry materials must be increased by 50% (approximately) to get the required quantity of concrete. The detailed calculations are given in Appendix 1.

Example 12.7

A concrete drive to a garage is to be 10 m long, 3 m wide and 150 mm thick. Calculate:

(a) The volume of concrete required to construct the drive

(b) The quantities of cement, sand and gravel required if 1:2:4 concrete is to be used.

Solution:

(a) Convert 150 mm into metres, as the units of all measurements should be the same:

$$150 \text{ mm} = \frac{150}{1000} = 0.15 \text{ m}$$

The shape of the drive is a rectangular prism:

$$\text{Volume of concrete required} = lwh$$
$$= 10 \times 3 \times 0.15$$
$$= \textbf{4.5 m}^3$$

(b) The quantities of dry materials to make 1 m^3 are (from Appendix 1): cement = 0.198 m^3; sand = 0.397 m^3; gravel = 0.8 m^3; water = 109 litres. The quantities of these materials to make 4.5 m^3 are:

Cement = 4.5 × 0.198 = **0.891 m^3**

Sand = 4.5 × 0.397 = **1.787 m^3**

Gravel = 4.5 × 0.8 = **3.60 m^3**

Water = 4.5 × 109 = **491 litres**

Example 12.8

Figure 12.9 shows the plan of the proposed extension to a building. Find:

(a) the volume of the earth to be excavated.

(b) the volume of concrete required to construct the deep strip foundation.

Solution:

Figure 12.9 shows the sectional details of the deep strip foundation. The foundation may be considered to be a rectangular prism; hence its volume will be the product of the plan area and the depth. The width of the trench will be equal to the width of the foundation, i.e. 500 mm (or 0.5 m). The foundation plan, shown in Figure 12.10 has been divided into parts A, B and C to simplify the calculations.

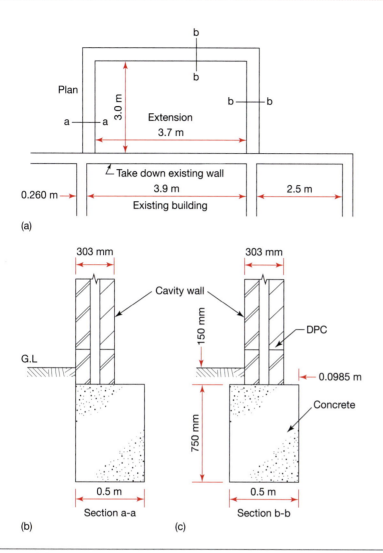

Figure 12.9

The total cross-sectional area will be the sum of the areas of parts A, B and C:

Area of part A = 2.9015 × 0.5 = 1.45075 m²

Area of part B = 4.4045 × 0.5 = 2.20225 m²

Area of part C = Area of part A = 1.45075 m²

Total area = 1.45075 + 2.20225 + 1.45075 = 5.10375 m²

Depth of the trench = 900 mm or 0.9 m

Volume of earth = 5.10375 × 0.9 = **4.593 m³**

Depth of the foundation = 750 mm = 0.75 m

Volume of concrete = 5.10375 × 0.75 = **3.828 m³**

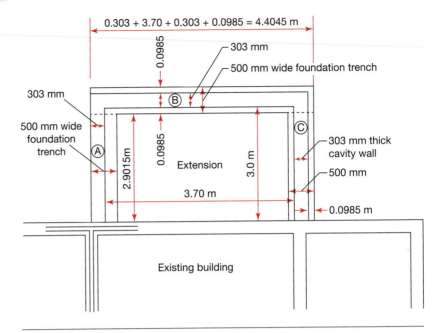

Figure 12.10

Exercise 12.1

The solutions to Exercise 12.1 can be found in Appendix 2.

1. Find the volume of a room measuring 3.24 m × 4.38 m × 2.46 m.

2. A cylinder has a diameter of 450 mm and a height of 950 mm. Find the volume of the cylinder (a) in cubic metres; (b) in litres. Note that 1 m³ is equal to 1000 litres.

3. A decorative feature on the front of a college building takes the form of a square-based pyramid of side 3.5 m, while its height is 6.0 m. Calculate its volume.

4. The spire of a church is in the form of a cone measuring 8 m high. If the diameter of its base is 3 m, what will be its volume?

5. Calculate the volume of concrete contained in the solid feature shown in Figure 12.11.

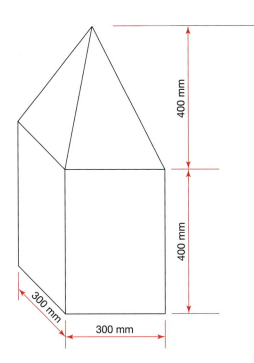

Figure 12.11

6. Figure 12.12 shows the dimensions of the roof of a building. Calculate the volume of the loft space.

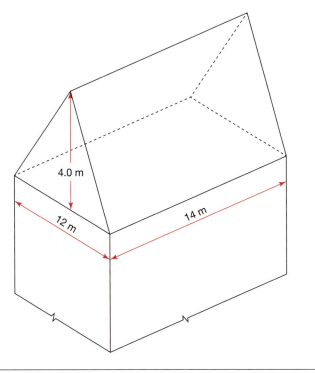

Figure 12.12

7. A rectangular culvert is to be used to enclose and protect a small river where it passes through an industrial estate. The culvert has external dimensions of 1.75 m by 1.25 m, and has a wall thickness of 150 mm. What will be the volume of concrete used to manufacture each linear metre of the culvert?

8. Figure 12.13 shows part of an embankment and its dimensions. What is its volume?

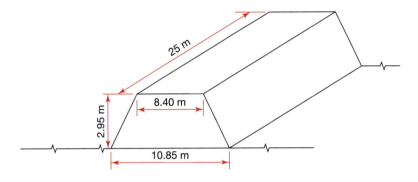

Figure 12.13

9. A water storage cistern measuring 2.4 m × 1.8 m × 1.2 m high is to be provided in a building.

   (a) Calculate the volume of water when the cistern is filled to 15 cm from the top rim.

   (b) Use the volume of water calculated in part (a) to determine the mass of water. The density of water is 1000 kg/m³.

   (c) Find the load (force) on the joists supporting the cistern. 1 kg = 9.8 newton (N).

10. A concrete drive to a garage is to be 8.5 m long, 3.5 m wide and 150 mm thick. Calculate:

   (a) the volume of concrete required to construct the drive

   (b) the quantities of cement, sand and gravel required if 1:2:4 concrete is to be used.

11. The plan of a building extension is shown in Figure 12.14.

   (a) Calculate the volume of soil to be excavated from the foundation trench. The depth of the trench is 900 mm.

   (b) If the soil bulks by 15%, find its volume after bulking.

   (c) Find the volume of concrete required to construct a 600 mm wide and 200 mm thick strip foundation.

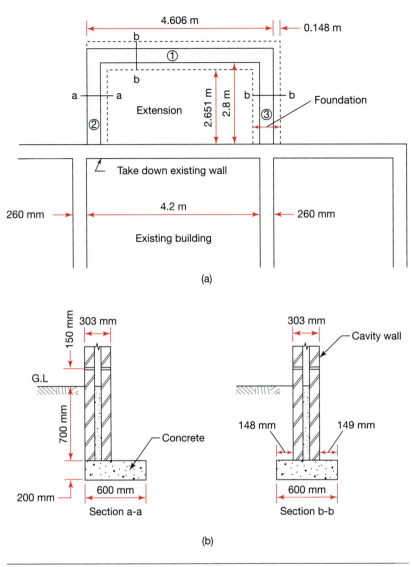

Figure 12.14

Answers to Exercise 12.1

1. 34.910 m³
2. (a) 0.15109 m³; (b) 151.09 litres
3. 24.5 m³
4. 18.85 m³
5. 48 000 000 mm³ or 0.048 m³
6. 336 m³

7. 0.81 m^3

8. 709.844 m^3

9. (a) 4.536 m^3; (b) 4536 kg; (c) 44 452.8 N

10. (a) 4.4625 m^3; (b) Cement = 0.884 m^3, sand = 1.772 m^3, gravel = 3.57 m^3, water = 486 litres

11. (a) 5.430 m^3; (b) 6.245 m^3; (c) 1.207 m^3

CHAPTER 13

Trigonometry (1)

Learning outcomes:

(a) Perform calculations to determine the sine, cosine and tangent of given angles and vice versa

(b) Perform calculations to find the angle of pitch and headroom of a flight of stairs

(c) Find the true length of common and hip rafters and the area of a roof

13.1 Introduction

The word trigonometry comes from two Greek words, *trigonon* (triangle) and *metria* (measure). Trigonometry may be defined as a branch of mathematics that deals with the study of relationships between the sides and the angles of triangles (Δ). The complex history of the term 'sine' reveals that the origins of trigonometry trace to the ancient cultures of Egyptian, Babylonian, Greek and Indus Valley civilisations. Originally, the use of trigonometry became popular as developments started to take place in astronomy. To calculate the position of the planets the astronomers used concepts we now refer to as trigonometry.

The use of trigonometry is not just limited to mathematics. It is also used in physics, land surveying, engineering, satellite navigation and other applications.

13.2 The trigonometrical ratios

Consider a right-angled triangle CAB, shown in Figure 13.1

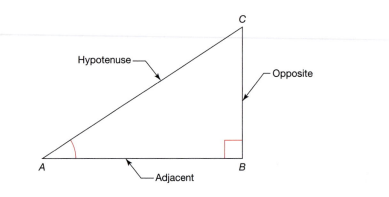

Figure 13.1

Angle *CBA* (∠*B*) is a right angle, i.e. 90°. If ∠*A* is being considered, side *BC* is called the 'opposite side' or opposite. Side *AC*, which is the longest side, is called the 'hypotenuse'. The third side, *AB*, is called the 'adjacent side' or adjacent and is common to ∠*A* as well as the right angle.

The ratios of the sides of a right-angled triangle, called the 'trigonometrical ratios' are sine, cosine and tangent. In △*CAB* (Figure 13.1):

$$\frac{\text{opposite}}{\text{hypotenuse}} = \frac{BC}{AC} = \text{sine of } \angle A \text{ or sin } A$$

$$\frac{\text{adjacent}}{\text{hypotenuse}} = \frac{AB}{AC} = \text{cosine of } \angle A \text{ or cos } A$$

$$\frac{\text{opposite}}{\text{adjacent}} = \frac{BC}{AB} = \text{tangent of } \angle A \text{ or tan } A$$

It is also known that $\tan A = \frac{\sin A}{\cos A}$. If △*CAB* is rotated clockwise through 90°, we have △*ABC*, as shown in Figure 13.2. If ∠*C* is being considered, then *AB* is the opposite side. *AC* and *BC* are the hypotenuse and adjacent side, respectively.

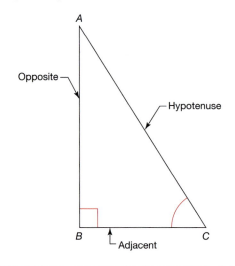

Figure 13.2

The trigonometric ratios in this case are:

$$\frac{\text{opposite}}{\text{hypotenuse}} = \frac{AB}{AC} = \text{sine of } \angle C \text{ or } \sin C$$

$$\frac{\text{adjacent}}{\text{hypotenuse}} = \frac{BC}{AC} = \text{cosine of } \angle C \text{ or } \cos C$$

$$\frac{\text{opposite}}{\text{adjacent}} = \frac{AB}{BC} = \text{tangent of } \angle C \text{ or } \tan C$$

13.3 Trigonometric ratios for 30°, 45°, 60°

The 30°, 60°, 90° and 45°, 45°, 90° triangles are shown in Figure 13.3

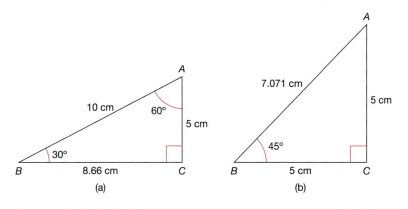

Figure 13.3

Figure 13.3a shows a 30°, 60°, 90° triangle whose sides are in the ratio 1:2:√3 (or 1:2:1.732). If side AC = 5 cm, then AB, being two times AC, is equal to 10 cm. Using Pythagoras' Theorem, BC = 8.66 cm.

$$\sin 30° = \frac{\text{opposite}}{\text{hypotenuse}} = \frac{AC}{AB} = \frac{5}{10} = 0.5$$

$$\cos 30° = \frac{\text{adjacent}}{\text{hypotenuse}} = \frac{BC}{AB} = \frac{8.66}{10} = 0.866$$

$$\tan 30° = \frac{\text{opposite}}{\text{adjacent}} = \frac{AC}{BC} = \frac{5}{8.66} = 0.577$$

Figure 13.3b shows a 45°, 45°, 90° triangle whose sides are in the ratio 1:1:√2 or (1:1:1.4142). Side $BC = AC$ = 5 cm, and AB = 7.071 cm. Following the above process:

$$\sin 45° = \frac{5}{7.071} = 0.707$$

$$\cos 45° = \frac{5}{7.071} = 0.707$$

$$\tan 45° = \frac{5}{5} = 1.0$$

A summary of the above values (correct to 3 d.p.) and the values for other angles which can be determined using the above procedure are given in Table 13.1

Table 13.1

	Angle				
	0°	30°	45°	60°	90°
sin	0	0.5	0.707	0.866	1
cos	1	0.866	0.707	0.5	0
tan	0	0.577	1	1.732	Infinity

Example 13.1

Determine angles A and B of $\triangle ABC$ shown in Figure 13.4.

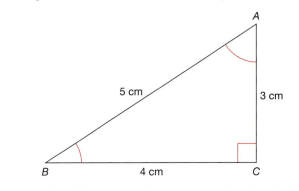

Figure 13.4

Solution:

With reference to $\angle B$, label the sides as adjacent, opposite and hypotenuse:

BC = adjacent; AC = opposite; AB = hypotenuse

As all three sides of the triangle are known, any trigonometric ratio can be used to find $\angle B$.

$$\sin B = \frac{\text{opposite}}{\text{hypotenuse}} = \frac{AC}{AB} = \frac{3}{5} = 0.6$$

Using a scientific calculator:

$\angle B = \sin^{-1} 0.6 = $ **36.87° or 36°52′12″**

(Refer to Chapter 1 for instructions on the use of a scientific calculator.)

$\angle A$ can be calculated by two methods.

Method 1
The sum of all angles of a triangle = 180°
Therefore, $\angle A = 180° - \angle B - \angle C$

$= 180° - 36°52′12″ - 90° = $ **53°07′48″**

Method 2

With reference to $\angle A$, BC = opposite side

$\qquad\qquad\qquad AB$ = hypotenuse

$\qquad\qquad\qquad AC$ = adjacent side

$$\cos A = \frac{\text{adjacent}}{\text{hypotenuse}} = \frac{AC}{AB} = \frac{3}{5} = 0.6$$

$\angle A = \cos^{-1} 0.6 = 53.13°$ or **53°07′48″**

Example 13.2

Find side AB of the triangle shown in Figure 13.5.

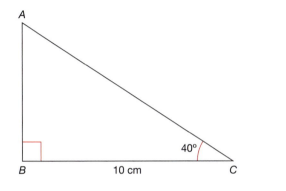

Figure 13.5

Solution:

With reference to $\angle C$ in the above triangle:

$\quad AB$ = opposite side (to be determined)

$\quad BC$ = adjacent side = 10 cm

The trigonometric ratio involving sides AB and BC is the tangent ratio:

$$\tan C = \frac{\text{opposite}}{\text{adjacent}} = \frac{AB}{BC}$$

$$\tan 40° = \frac{AB}{10}$$

Transposing, $\tan 40° \times 10 = AB$

Therefore $AB = 0.839 \times 10 =$ **8.39 cm**   ($\tan 40° = 0.839$)

Example 13.3

A flat roof of 3 m span has a fall of 75 mm. Find the pitch of the roof.

Solution:

The cross-section of the roof is shown in Figure 13.6:

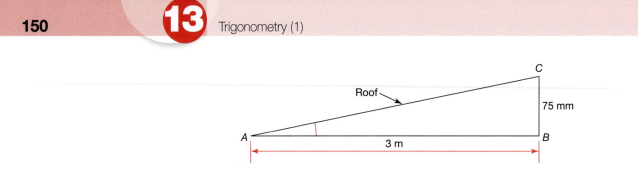

Figure 13.6

The pitch of a roof is the angle that it makes with the horizontal. In this example $\angle A$ is the pitch of the roof. Convert 3 m into millimetres (or 75 mm into metres) to have both dimensions in the same units.

$$3 \text{ m} = 3.0 \times 1000 = 3000 \text{ mm} \quad (1 \text{ m} = 1000 \text{ mm})$$

As the opposite and adjacent sides are known, the tangent ratio will be used:

$$\tan A = \frac{\text{opposite}}{\text{adjacent}} = \frac{BC}{AB}$$

$$= \frac{75}{3000} = 0.025$$

$$\angle A \text{ or Pitch} = \tan^{-1} 0.025$$

$$= \mathbf{1.432° \text{ or } 1°25'56''}$$

Example 13.4

The gradient of a road is 1 in 5. Find the angle that the road makes with the horizontal.

Solution:

Figure 13.7 shows the slope of the road:

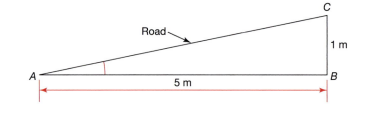

Figure 13.7

Gradient of 1 in 5 means that for every 5 m of horizontal distance there is a rise or fall of 1 m. If AB represents the horizontal distance of 5 m, the rise of 1 m is represented by BC. To calculate $\angle A$, the tangent ratio will be used as the opposite and adjacent sides are known

$$\tan A = \frac{\text{opposite}}{\text{adjacent}} = \frac{BC}{AB} = \frac{1}{5} = 0.2$$

$$\therefore \angle A = \tan^{-1} 0.2 = \mathbf{11.31° \text{ or } 11°18'36''}$$

13.4 Angles of elevation and depression

If an observer looks at an object that is higher than him, the angle between the line of sight and the horizontal is called an angle of elevation (Figure 13.8a). When the observer is higher than the object, the angle, as shown in Figure 13.8b, is called an angle of depression.

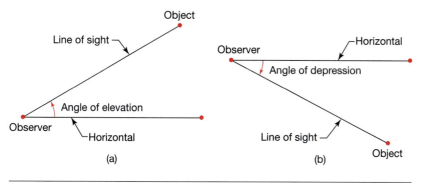

Figure 13.8

Example 13.5

A surveyor, 100 m from a building, measures the angle of elevation to the top of a building to be 40° (Figure 13.9). If the height of the instrument is 1.400 m and the ground between the surveyor and the building is level, find the height of the building.

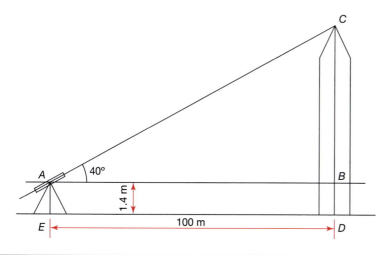

Figure 13.9

Solution:

Height of the building $= CD = BC + BD$

Height of the instrument $= AE = BD = 1.4$ m

$$\tan 40° = \frac{BC}{AB}$$

Transposing, $BC = AB \times \tan 40°$

$$= 100 \times 0.8391 = 83.91 \text{ m}$$

Height of the building, $CD = BC + BD$

$$= 83.91 + 1.400 = \mathbf{85.310 \text{ m}}$$

Example 13.6

Jane, standing on the fifteenth floor of a building, looks at her car parked on the nearby road. Find the angle of depression if the other details are as shown in Figure 13.10.

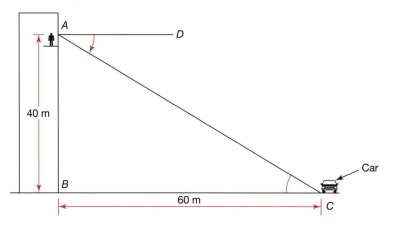

Figure 13.10

Solution:

Assuming the building is vertical and the ground is level, horizontal line AD is parallel to BC. Therefore, the angle of depression $\angle A = \angle C$. To calculate $\angle C$, consider $\triangle ABC$: $\angle B = 90°$, AB = opposite side, and BC = adjacent side:

$$\tan C = \frac{\text{opposite}}{\text{adjacent}} = \frac{AB}{BC} = \frac{40}{60} = 0.6667$$

$$\therefore \angle C = \tan^{-1} 0.6667 = \mathbf{33°41'24''}$$

13.5 Stairs

Stairs in buildings provide access from one floor to another. The Building Regulations give guidelines on their design for use in domestic and commercial buildings. According to the Building Regulations 2000:

1. The pitch (see Figure 13.11) of stairs in dwelling houses should not be more than 42°

2. Maximum rise (R) = 220 mm

3. Minimum going (G) = 220 mm

4. 2 × Rise + Going (or 2R + G) must be within 550 and 700 mm.
5. The headroom should not be less than 2.0 m.

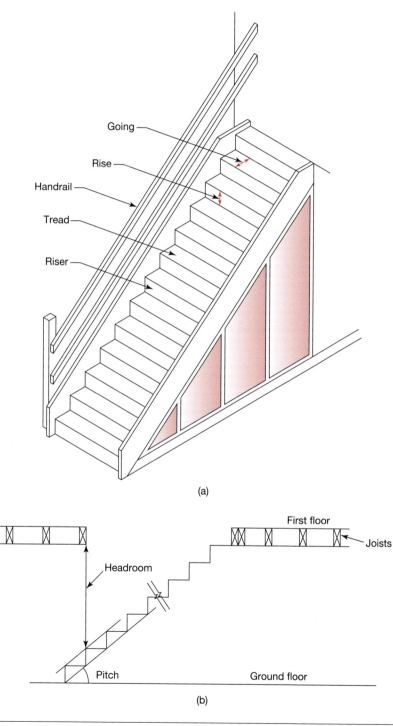

(a)

(b)

Figure 13.11

Example 13.7

A flight of stairs has 12 steps. Each step has a rise of 210 mm and going equal to 230 mm.

(a) Find the pitch of the stairs

(b) If the pitch exceeds the limit set by the Building Regulations, find the satisfactory dimensions of the rise and going.

Solution:

(a) Figure 13.12 shows a portion of the stairs

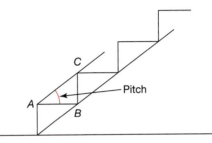

Figure 13.12

Join points A and C. In $\triangle CAB$

AB = Going = 230 mm

BC = Rise = 210 mm

$\angle B = 90°$

$\angle A$ = The pitch of the stairs (to be calculated).

With reference to $\angle A$, in $\triangle CAB$:

Side BC = opposite side, and side AB = adjacent side

$$\tan A = \frac{\text{opposite}}{\text{adjacent}} = \frac{BC}{AB} = \frac{210}{230} = 0.913$$

$\therefore \angle A = \tan^{-1} 0.913 = \textbf{42.4°}$

The current Building Regulations stipulate that the maximum pitch of a private stair should not be more than 42°.

(b) To reduce the pitch, increase the going (AB) to, say, 235 mm

$$\tan A = \frac{\text{opposite}}{\text{adjacent}} = \frac{BC}{AB} = \frac{210}{235} = 0.8936$$

$\therefore \angle A = \tan^{-1} 0.8936 = \textbf{41.8°} < 42°$, satisfactory

(Note: the pitch may also be reduced by decreasing the rise to (say) 205 mm.)

$2R + G = 2 \times 210 + 235$

$= 420 + 235 = 655$

which is more than (>) 550, but less than (<) 700, i.e. within the permissible limits.

Example 13.8

The floor to floor height in a house is 2.575 m and the space to be used for providing the stairs is shown in Figure 13.13. Design a staircase that satisfies the requirements of the Building Regulations.

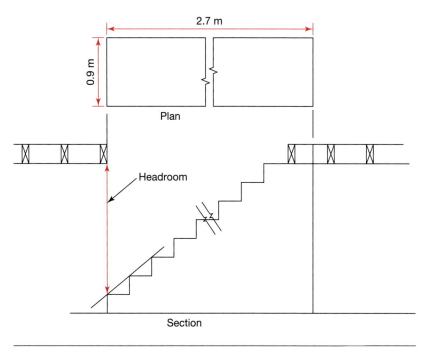

Figure 13.13

Solution:

Distance between two floors (height) = 2575 mm. Table 13.2 shows the number of risers and the rise of each step

Table 13.2

Number of risers	Rise of each step
10	2575 ÷ 10 = 257.5 mm > 220 mm (unsatisfactory)
11	2575 ÷ 11 = 234.09 mm > 220 mm (unsatisfactory)
12	2575 ÷ 12 = 214.58 mm (satisfactory)
13	2575 ÷ 13 = 198.08 mm (satisfactory)

A staircase with 12 or 13 risers can be provided

Number of risers = 12:

Number of treads = 11; (one less than the number of risers)

Rise of each step = 214.58 mm

Going of each step = 2700 ÷ 11 = 245.45 mm

In Figure 13.14, join points *A* and *C* to form Δ*CAB*. In Δ*CAB*:

∠*A* = Pitch of the stair

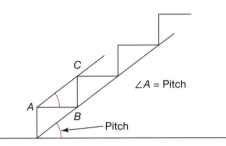

Figure 13.14

AB = Going = 245.45 mm

BC = Rise = 214.58 mm

$$\tan A = \frac{\text{opposite}}{\text{adjacent}} = \frac{BC}{AB} = \frac{214.58}{245.45}$$

$$\text{Pitch} = \angle A = \tan^{-1}\frac{214.58}{245.45}$$

∴ ∠*A* = tan⁻¹ 0.874 = 41.16° < 42°, satisfactory

Number of risers = 13:

Rise of each step = 198.08 mm

Number of steps = 12

Going of each step = 2700 ÷ 12 = 225 mm

$$\text{Pitch} = \angle A = \tan^{-1}\frac{198.08}{225}$$

∴ ∠*A* = tan⁻¹ 0.88 = 41.36° < 42°, satisfactory

Headroom

Assuming the depth of first-floor joists to be 250 mm, the total depth of first-floor construction is 285 mm (approx.).

Floor to ceiling height = 2.575 – 0.285 = 2.290 m

From Figure 13.13, the headroom = 2.290 – 0.215 m

= 2.075 m

The headroom is satisfactory as it is more than 2.0 m.

13.6 Roofs

Some of the terms associated with roof construction are shown in Figure 13.15.

The gable end is a vertical wall right up to the ridge, but in a hipped roof the vertical wall does not extend beyond the eaves. The hipped end normally slopes at the same angle to the main roof, and is tiled like the rest of the roof.

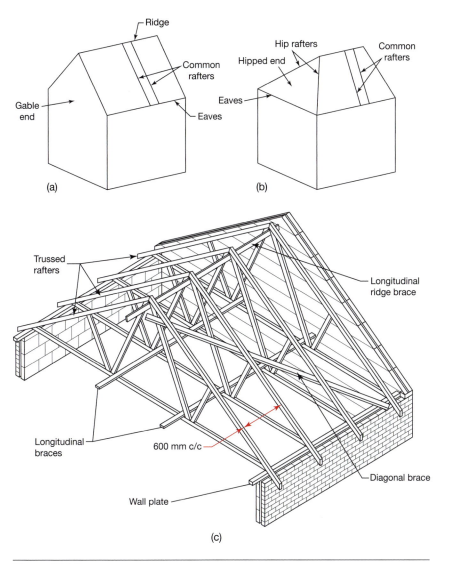

Figure 13.15

The true length of a rafter can be seen and calculated from some of the views, as shown in Example 13.9.

Example 13.9

The roof shown in Figure 13.16 has a height of 4 m and a span of 10 m. Calculate:

(a) Pitch of the roof

(b) True lengths of common rafters

(c) Surface area of the roof

(d) The number of single lap tiles required to cover the roof.

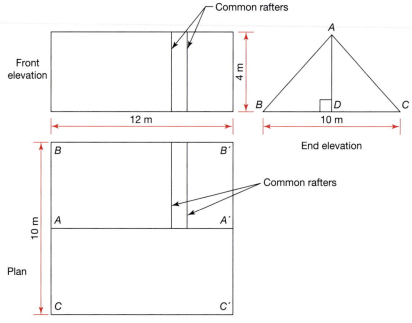

Figure 13.16

Solution:

(a) As the roof is symmetrical, $BD = DC$ and $AB = AC$. With reference to $\angle B$ in the right-angled triangle ABD:

AD = opposite side = 4 m; BD = adjacent side = 5 m

$$\tan B = \frac{AD}{BD} = \frac{4}{5} = 0.8$$

$\therefore \ \angle B = \tan^{-1} 0.8 = 38.66°$

The pitch of the roof is **38.66°**

(b) The plan and the front elevation do not give the true length of a rafter. From the front elevation a common rafter appears to be 4 m long and from the plan it appears to be 5 m. Both are incorrect. The true length of a common rafter is equal to AB or AC in $\triangle ABC$

$$\cos B = \frac{BD}{AB} \text{ or } AB = \frac{BD}{\cos B}$$

$$= \frac{5}{\cos 38.66°} = \frac{5}{0.7809} = 6.403 \text{ m}$$

$\therefore$ True length of the common rafter = **6.403 m**

(This can also be calculated by using Pythagoras' Theorem.)

(c) Again, the plan area of the roof surface does not give the correct value. Referring to Figure 13.17:

Surface area of the roof = 2 × Area of surface $A\,A'\,C'\,C$

$AC = A'C' = 6.403$ m

Therefore,

surface area of the roof $= 2 \times (AA' \times AC)$

$\qquad\qquad\qquad\qquad\quad = 2 \times (12 \times 6.403)$

$\qquad\qquad\qquad\qquad\quad =$ **153.672 m²**

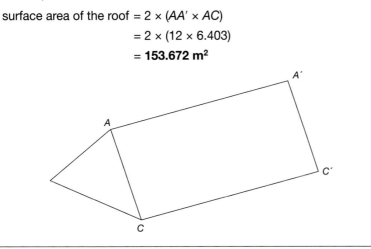

Figure 13.17

(d) The size of a trough-type single lap roof tile is 330 mm × 413 mm.

The exposed area of each tile is:

292 mm × 338 mm = 98 696 mm²

$\qquad\qquad\qquad\qquad\quad = 9.8696 \times 10^{-2}$ m²

(Alternatively: 0.292 m × 0.338 m = 0.0987 m²)

Number of tiles $= \dfrac{\text{Surface area of roof}}{\text{Area of a tile}}$

$\qquad\qquad\quad = \dfrac{153.672}{9.8696 \times 10^{-2}}$

$\qquad\qquad\quad = 1557.02$ or **1558**

Example 13.10

The pitch of a 14 m long hipped roof is 45°. If other dimensions are as shown in Figure 13.18, find:

(a) Height of the roof

(b) Length of common rafter XZ

(c) True length of hip rafter DA.

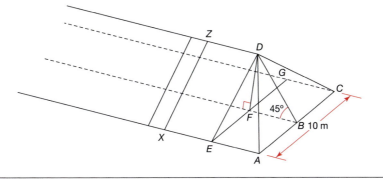

Figure 13.18

Solution:

(a) The height of the roof is the same as the height *DF*, a vertical line that makes an angle of 90° with the horizontal. It can be determined from the right-angled triangle *DEF*.

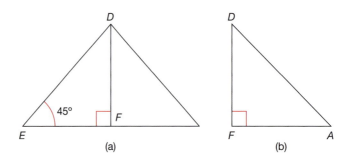

(a)   (b)

Figure 13.19

In Δ*DEF* (see Figure 13.19a), ∠*E* = 45° and *EF* = 5 m

$$\frac{DF}{EF} = \frac{\text{opposite}}{\text{adjacent}} = \tan 45°$$

$$\frac{DF}{5} = 1$$

Transposing, *DF* = 1 × 5 = **5 m**

Hence the height of the roof is **5 m**

(b) Common rafter *XZ* = *DE*; *DE* can also be determined from Δ*DEF*

$$\frac{DF}{DE} = \frac{\text{opposite}}{\text{hypotenuse}} = \sin 45°$$

$$\frac{5}{DE} = 0.7071$$

$$DE = \frac{5}{0.7071} = 7.071 \text{ m}$$

True length of rafter *XZ* = **7.071 m**

(c) To find the length of rafter *DA*, consider the right-angled triangle *DFA* (Figure 13.19b). *DF* = 5 m; but this information is not enough to find *DA*. We need to find another side of Δ*DFA* to do further calculations.

    ABFE is a square with each side equal to 5 m. *FA* is the diagonal of the square, which can be determined using Pythagoras' Theorem:

$$FA = \sqrt{AB^2 + BF^2} = \sqrt{5^2 + 5^2} = \sqrt{25 + 25} = \sqrt{50} = 7.071 \text{ m}$$

$$\tan A = \frac{DF}{FA} = \frac{\text{opposite}}{\text{adjacent}} = \frac{5}{7.071} = 0.7071$$

$$\therefore \angle A = \tan^{-1} 0.7071 = 35.26°$$

$$\frac{DF}{DA} = \frac{\text{opposite}}{\text{hypotenuse}} = \sin A$$

$$\frac{5}{DA} = \sin 35.26°$$

Transposing, $5 = DA \times \sin 35.26°$

or $DA = \dfrac{5}{\sin 35.26°} = \dfrac{5}{0.577} = 8.660 \text{ m}$

Hip rafter DA is **8.660 m long.**

13.7 Excavations and embankments

Trench and pit excavations are required for the construction of foundations and other processes. The excavations can have vertical or battered (sloping) sides. Similarly, embankments which are quite commonly used in the construction of roads have battered sides.

Trigonometric ratios can be used to determine the volume of soil either removed from an excavation or required to construct an embankment. The following example is based on an excavation, but the same method can also be used to perform calculations in the case of an embankment.

Example 13.11

A 10 m long trench has battered sides which make angles of 40° with the horizontal. If the base of the trench is 1.0 m wide and depth 1.5 m, find the volume of the soil to be excavated.

Solution:

Volume of the excavated soil = cross-sectional area of trench × length
The cross-sectional area can be found either by treating the trench as a trapezium or by dividing the cross-section into three parts, i.e. two triangles and a rectangle (see Figure 13.20). The latter approach will be used here.

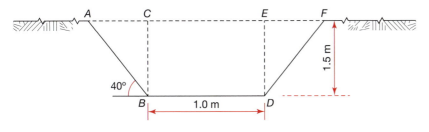

Figure 13.20

Cross-sectional area of the trench = area of △ABC + area of rectangle BCED + area of △DEF

Area of △ABC = area of △DEF

∴ Cross-sectional area of the trench = 2 × area of △ABC + area of rectangle BCED

In $\triangle ABC$, $\angle ABC = 90° - 40° = 50°$; AC = opposite side; BC = adjacent side

$$\frac{AC}{BC} = \frac{opposite}{adjacent} = \tan 50°$$

$$\frac{AC}{1.5} = 1.192$$

$$AC = 1.5 \times 1.192 = 1.788 \text{ m}$$

Area of $\triangle ABC$ = area of $\triangle DEF = \dfrac{base\ (BC) \times height\ (AC)}{2}$

$$= \frac{1.5 \times 1.788}{2} = 1.341 \text{ m}^2$$

Area of rectangle $BCED = 1.0 \times 1.5 = 1.5 \text{ m}^2$

Total cross-sectional area = 1.341 + 1.5 + 1.341 = 4.182 m^2

Volume of the trench = cross-sectional area × length = 4.182 × 10

= 41.82 m^3

Exercise 13.1

The solutions to Exercise 13.1 can be found in Appendix 2.

1. (a) Use a calculator to find the values of:
    (i) sin 30°20′35″
    (ii) cos 50°10′30″
    (iii) tan 40°55′05″.

   (b) Use a calculator to find the angles in degrees/minutes/seconds given that:
    (i) sine of the angle is 0.523
    (ii) cosine of the angle is 0.981
    (iii) tangent of the angle is 0.638.

2. In a right-angled triangle ABC (Figure 13.21) find the length of sides AC and BC if AB is 15 cm long.

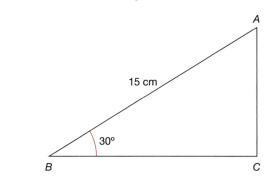

Figure 13.21

3. A surveyor wants to find the width of a river and stands on one bank at point C, directly opposite a building (B), as shown in Figure 13.22. He walks 60 m along the river bank to point A. If angle $BAC = 60°35'30''$, find the width of the river.

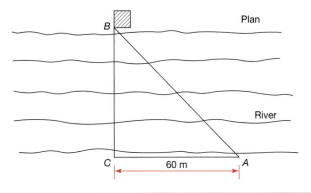

Figure 13.22

4. A tall structure is stabilised by five steel cables. The upper ends of the cables hold the top of the building and the lower ends are anchored into a concrete base. If the length of each cable is 35 m and they make an angle of 60° with the ground, find:

   (a) the height of the building

   (b) the distance between the building and the anchored end of one of the cables.

5. The gradient of a road is 9%. Find the vertical rise/fall if the length of the road is 2450 m.

6. John has designed a flat roof for his house extension. The span of the roof is 3.5 m and the slope of the roof is 1 in 48. Find the pitch of the roof in degrees.

7. The dimensions of a building plot are as shown in Figure 13.23. Find the area of the plot if AC is parallel to BD.

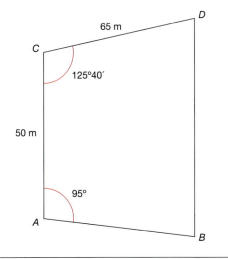

Figure 13.23

8. The pitch of a 15 m long hipped roof is 40°. If other dimensions are as shown in Figure 13.24, find the height of the roof and

   (a) true length of common rafter HJ

   (b) true length of hip rafter DA

   (c) area of the roof surface.

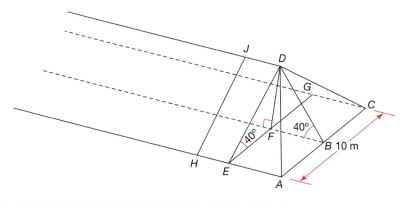

Figure 13.24

9. A 1.200 km long section of a road is provided on an embankment. The average cross-section of the embankment is as shown in Figure 13.25. Find the cross-sectional area and the volume of soil required to construct the embankment.

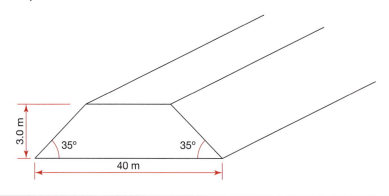

Figure 13.25

10. The flight of a domestic staircase is shown in Figure 13.26. Calculate the pitch of the staircase and check if it conforms to the Building Regulations. If necessary, select new dimensions of the rise and the going to satisfy the Building Regulations.

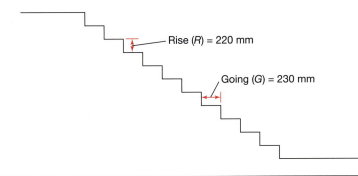

Figure 13.26

Answers to Exercise 13.1

1. (a) (i) 0.5052; (ii) 0.6404; (iii) 0.8668
   (b) (i) 31°32′ 1.3″; (ii) 11°11′ 12.4″; (iii) 32°32′ 16.4″
2. BC = 12.99 cm, AC = 7.5 cm
3. 106.447 m
4. (a) 30.311 m; (b) 17.5 m
5. 219.612 m
6. 1°11′ 36.56″
7. 3763.04 m^2
8. (a) HJ = 6.527 m, (b) DA = 8.222 m, (c) area = 195.81 m^2
9. Cross-section area = 107.147 m^2, volume = 128 576.4 m^3
10. rise = 201.67 mm, going = 224 mm

CHAPTER 14

Setting out

Learning outcomes:

(a) Set out a simple building

(b) Prepare data (ordinate lengths) for setting out circular curves

(c) Check that the corners of a building are square

(d) Prepare data for setting out elliptical arches

14.1 Introduction

'Setting out' is an expression used in the construction industry to cover the general measurement and control of the horizontal and vertical positions of buildings, roads and drains (and other features) on a building or civil engineering site. Some of the curvilinear shapes which are set out by builders and civil engineers include bay windows, circular brickwork, arches and curves in roads. In this chapter calculations required for setting out building sites, bay windows, curved brickwork and arches, and a brief procedure for utilising these calculations, are given.

14.2 Setting out a simple building site

Figure 14.1 shows a building site containing two blocks of factory units and one three-storey office block. When a site is originally surveyed for any construction work it is usual for some points (often referred to as stations) to be made so that they are permanent features and can be found at a later date. On Figure 14.1 two points, A and B, are such stations. Now, it is quite easy to see that on the drawing a pair of compasses could be used with the point in station A and the drawing tip set to 50.7 m (or 50 700 mm as shown on the drawing), to describe an arc near the lower corner of the three-storey block. Similarly, with the compass point in B another arc could be drawn with the compass set at 81.4 m. Where the arcs cross will give the position of the lower corner of the office block. A similar method would be used on-site, measuring the distances either with a steel or other type of measuring tape, or with an electronic distance measuring (EDM) device.

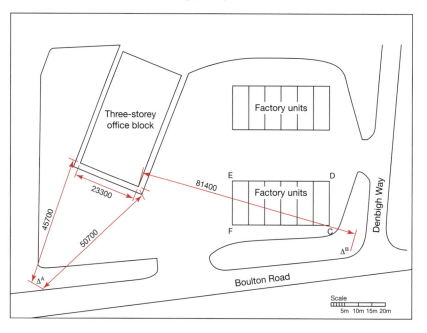

Figure 14.1

Having established the lower corner, a similar method would be used to find the next corner across the width of the building. The width is 23.3 m and its distance from *A* is 45.7 m. These distances would be set from the first corner and from station *A*. Having established one side and two corners of the building, it is easy to see that the other two corners may be set up in two ways: the first is to set up right-angles from each established corner and then measure the distances to the other corners; the second is to calculate the diagonals and find the corner where each measured diagonal intersects with the measured side. It is likely that both methods would be used: one to check the other. The right-angles could be set up using the 3:4:5 method (see Figure 14.2a), or with a cross-staff, or with a theodolite.

It is not usually acceptable to scale dimensions from a drawing: important dimensions should be stated but the architect should make certain that there is sufficient information in the form of angles and/or dimensions to enable the position of the building to be determined.

When the positions of the corners are established they are marked by driving wooden pegs into the ground. These measure about 50 mm × 50 mm and are usually in the region of 450 mm to 600 mm long. Once the pegs are driven in, the corner positions are checked a second time and marked more accurately by driving a nail into the top of the peg. However, when the trenches are excavated these pegs would be dug out by the excavator so profiles are used, two per corner (see Figures 14.2 and 14.3) so that the exact positions of the corners of the building can be

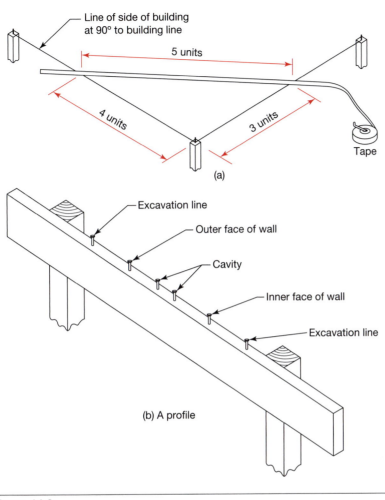

Line of side of building at 90° to building line

5 units

4 units

3 units

Tape

(a)

Excavation line

Outer face of wall

Cavity

Inner face of wall

Excavation line

(b) A profile

Figure 14.2

re-established after excavation. A profile is a piece of board about 25 mm by 125 mm nailed to two pegs set at least 2 m from the trench. A typical profile is shown in Figure 14.2b.

The nails along the top of the board represent the positions of the excavation lines, the outer and inner leaves of the wall and the cavity. Strings, or lines, are stretched from one profile to another so that the corners and faces of the brickwork can easily be re-established.

14.3 Bay windows and curved brickwork

There may be many methods available for setting out bay windows and curved brickwork, but here a method in which ordinates are used will be explained. This method requires the calculation of a number of ordinates, which later are measured off the base line (on-site). By joining the ends of these ordinates, a curve is obtained.

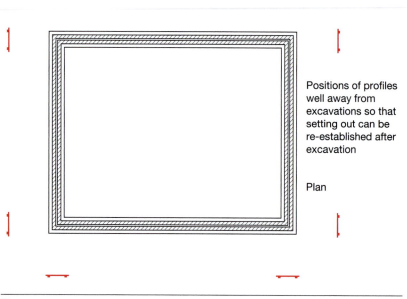

Positions of profiles well away from excavations so that setting out can be re-established after excavation

Plan

Figure 14.3

Figure 14.4a shows a circular curve with a span of 2.1 m and a rise of 0.5 m. To set out this curve, a number of ordinates are required at suitable intervals, but first of all it is necessary to calculate the radius of the curve. Figure 14.4b shows the circle of which curve $AD'G$ is a part. OA, OG and OD' are the radii, AD is 1.05 m (half the span) and DD' is the rise of 0.5 m:

OA and OD' are denoted as R and OD as r

$OD' = OD + DD'$

or $R = r + 0.5$ (1)

Also $r^2 = R^2 - (1.05)^2$

$\qquad = (r + 0.5)^2 - (1.05)^2$

$\qquad = r^2 + r + 0.25 - 1.1025$

$r^2 - r^2 = r + 0.25 - 1.1025$

$1.1025 - 0.25 = r$

$\therefore r = 0.8525$ or $OD = 0.8525$ m or 0.853 m (to 3 d.p.)

From equation 1,

$R = r + 0.5$

$\qquad = 0.8525 + 0.5 = 1.3525$ m or $= 1.353$ m (to 3 d.p.)

The above procedure may be generalised in the form of the following formula:

$$R = \frac{\left(\dfrac{\text{span}}{2}\right)^2 + (\text{rise})^2}{2 \times \text{rise}}$$

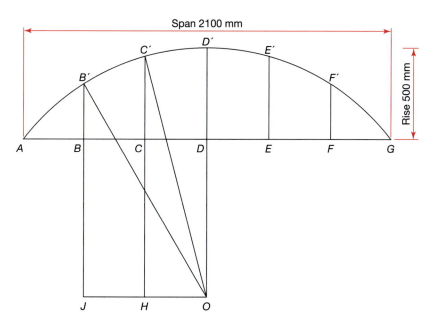

(a)

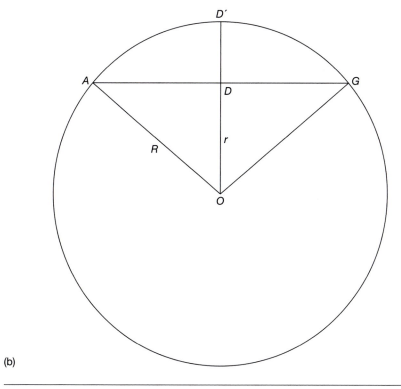

(b)

Figure 14.4

Other methods or formulae are available to work out this radius. Now divide the base line into an even number of equal parts (six or more). Here we have used six. The next stage is to determine ordinates CC' and BB'. Consider triangle $OC'H$ from Figure 14.4a.

Using Pythagoras' Theorem, $(C'H)^2 = (OC')^2 - (OH)^2$

$$C'H = \sqrt{(OC')^2 - (OH)^2}$$

$$C'H = \sqrt{(1.353)^2 - (0.350)^2}$$

$$C'H = \sqrt{1.708}$$

$$C'H = 1.307 \text{ m}$$

$$CC' = C'H - CH$$

$$= 1.307 - 0.853 = 0.454 \text{ m}$$

Similarly, in triangle $OB'J$:

$$B'J = \sqrt{(OB')^2 - (OJ)^2}$$

$$B'J = \sqrt{(1.353)^2 - (0.70)^2}$$

$$B'J = \sqrt{1.341} = 1.158 \text{ m}$$

$$B'B = B'J - BJ$$

$$BB' = 1.158 - 0.853 = 0.305 \text{ m.}$$

Due to the symmetry of the diagram:

$$EE' = CC' = 0.454 \text{ m}$$

$$FF' = BB' = 0.305 \text{ m}$$

From the established base line AG, measure and mark ordinates BB', CC', DD', EE' and FF' on the ground and join the points A, B', C', D' ,etc. with a smooth curve.

This method can also be used to set out segmental arches.

14.4 Checking a building for square corners

After setting out a building it is important to check that all corners are square (make 90° angles). Example 14.1 shows the procedure of checking the corners.

Example 14.1

A building is set out using a builder's square (see Figure 14.5). Show how the builder can check that the corners are square.

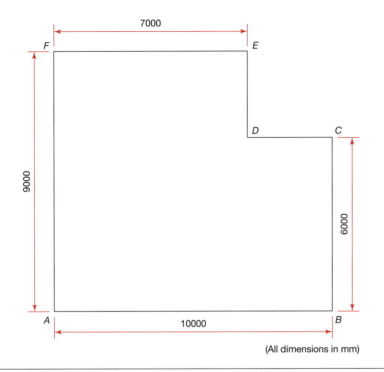

Figure 14.5

Solution:

Divide the area into rectangles, as shown in Figure 14.6

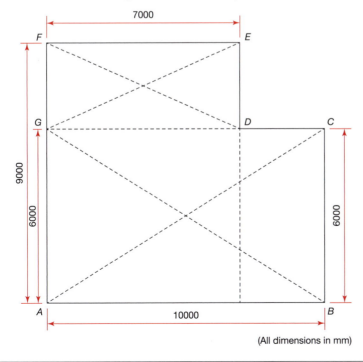

Figure 14.6

(a) Diagonals *AC* and *BG* should be equal.

$$AC = \sqrt{(AB)^2 + (BC)^2}$$
$$AC = \sqrt{10^2 + 6^2}$$
$$AC = \sqrt{100 + 36}$$
$$AC = \sqrt{136} = \textbf{11.662 m}. \text{ Diagonal } BG \text{ should also be } \textbf{11.662 m}$$

(b) Diagonals *GE* and *DF* should also be equal.

$$GE = \sqrt{(GD)^2 + (DE)^2}$$
$$GE = \sqrt{7^2 + 3^2}$$
$$GE = \sqrt{49 + 9}$$
$$GE = \sqrt{58} = \textbf{7.616 m}$$

(c) Additionally, *BF* and *AE* should be checked.

$$BF = \sqrt{AB^2 + AF^2}$$
$$BF = \sqrt{10^2 + 9^2}$$
$$BF = \sqrt{100 + 81}$$
$$BF = \sqrt{181} = \textbf{13.454 m}$$
$$AE = \sqrt{(AF)^2 + (FE)^2}$$
$$AE = \sqrt{9^2 + 7^2}$$
$$AE = \sqrt{81 + 49}$$
$$AE = \sqrt{130} = \textbf{11.402 m}$$

The builder should measure the diagonals *AC*, *BG*, *GE* and *DF*. *AC* and *BG* should be equal at 11.662 m long. Similarly, *GE* and *DF* should both be equal to 7.616 m. Additionally, *BF* and *AE* may be measured: they should both be as stated above.

14.5 Circular arches

Stone and brick arches have been in use in the construction of buildings and bridges for more than 3000 years. These are still used in buildings, though not so much for their structural purpose as their architectural appearance. Figure 14.7 shows the shapes of some of the arches which have been used in buildings, viaducts and aqueducts.

Because brick and stone in arches cannot support themselves before the mortar has gained enough strength, temporary supports known as centres are used, as shown in Figure 14.8. In order to construct a centre a number of ordinates are required at suitable intervals, as shown in Figures 14.9 and 14.10. The upper surface of the centre is shaped according to the required appearance of the arch.

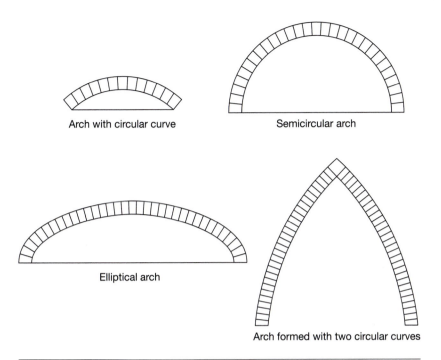

Arch with circular curve

Semicircular arch

Elliptical arch

Arch formed with two circular curves

Figure 14.7

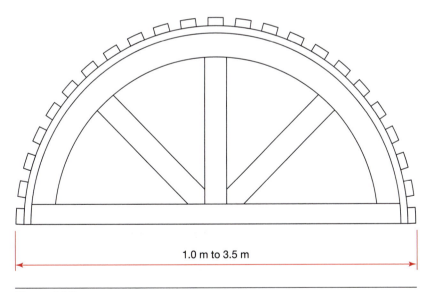

1.0 m to 3.5 m

Figure 14.8

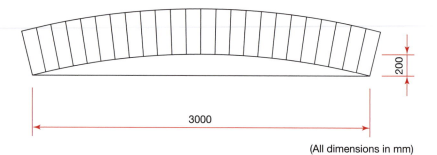

(All dimensions in mm)

Figure 14.9

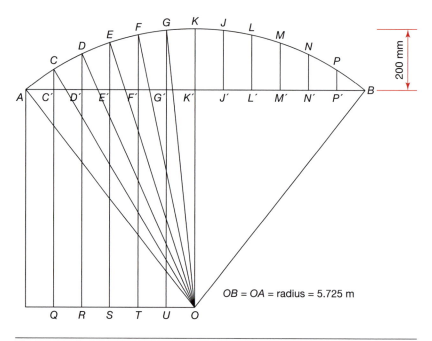

$OB = OA = $ radius $= 5.725$ m

Figure 14.10

Example 14.2

Calculate the ordinates required to set out the arch shown in Figure 14.9.

Solution:

To find the radius of the curve, use the following formula:

$$R = \frac{\left(\dfrac{\text{span}}{2}\right)^2 + (\text{rise})^2}{2 \times \text{rise}}$$

$$R = \frac{\left(\dfrac{3}{2}\right)^2 + (0.2)^2}{2 \times 0.2} = 5.725 \text{ m}$$

To find the lengths of the ordinate, use Pythagoras' Theorem (as explained in Section 14.3). Draw the outline of the arch as shown in Figure 14.10. Establish the centre O and divide AB into 12 equal parts, each 3.0 m ÷ 12 = 0.25 m apart.

To find ordinate GG' consider triangle OGU:

OG = radius = 5.725 m

OU = 0.250 m

$GU = \sqrt{(5.725)^2 - (0.25)^2}$

$GU = \sqrt{(32.776) - (0.0625)}$

$GU = \sqrt{32.713} = 5.720$ m

Ordinate $GG' = GU - G'U$

$GG' = 5.70 - 5.525 = \textbf{0.195 m}$

To find ordinate FF':

$FT = \sqrt{(FO)^2 - (OT)^2}$

$FT = \sqrt{(5.725)^2 - (0.5)^2}$

$FT = \sqrt{32.776 - 0.25} = 5.703$ m

Ordinate $FF' = FT - F'T$

$FF' = 5.703 - 5.525 = \textbf{0.178 m}$

To find ordinate EE':

$ES = \sqrt{(EO)^2 - (OS)^2}$

$ES = \sqrt{(5.725)^2 - (0.75)^2}$

$ES = 5.676$ m

Ordinate $EE' = ES - E'S$

$EE' = 5.676 - 5.525 = \textbf{0.151 m}$

To find ordinate DD':

$DR = \sqrt{(OD)^2 - (RO)^2}$

$DR = \sqrt{(5.725)^2 - (1.0)^2}$

$DR = 5.637$ m

$DD' = 5.637 - 5.525 = \textbf{0.112 m}$

To find ordinate CC'

$CQ = \sqrt{(CO)^2 - (QO)^2}$

$CQ = \sqrt{(5.725)^2 - (1.25)^2}$

$CQ = 5.587$ m

$CC' = 5.587 - 5.525 = \textbf{0.062 m}$

14.6 Elliptical arches

Example 14.3

Calculate the ordinates for setting out an elliptical arch having a span of 3.0 m and a rise of 0.45 m

Solution:

The horizontal and vertical axes of an ellipse are called the major and minor axes, respectively (see Figure 14.11a).

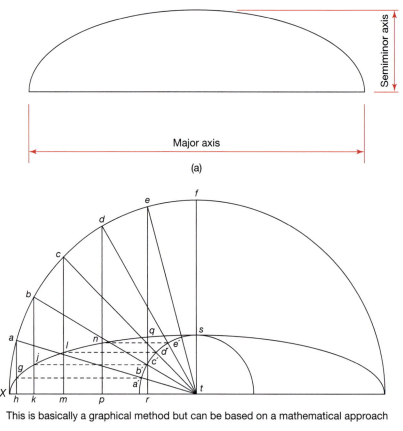

(a)

This is basically a graphical method but can be based on a mathematical approach

(b)

Figure 14.11

In this example the major axis = span = 3.0 m; the minor axis = 2 × rise = 2 × 0.45 = 0.90 m. Draw the two semicircles from the same centre, one with a radius of 1.5 m (half the major axis) and the other with 0.45 m (half the minor axis), as shown in Figure 14.11b.

Draw five or any other convenient number of radii *te*, *td*, *tc*, *tb* and *ta*. A smooth curve passing through points *x*, *g*, *j*, *l*, *n*, *q* and *s* will yield part of the ellipse. Because the ellipse is symmetrical the calculated

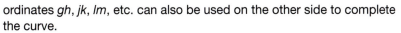

ordinates *gh*, *jk*, *lm*, etc. can also be used on the other side to complete the curve.

To locate the points *g*, *j*, *l*, *n* and *q* mathematically, draw vertical lines *er*, *dp*, *cm*, *bk* and *ah*. Now points *a*, *b*, *c*, *d*, *e* and *f* divide the quarter-circle into six equal parts so that the angle that each part makes at the centre is 15°.

In triangle *ert*:

$er/te = \sin \angle rte$

$\therefore er = te \times \sin\angle rte$

$\qquad = 1500 \times \sin 75°$

$\qquad = 1500 \times 0.966 = 1448.9$ mm

In triangle *eqe′*:

$eq/e′e = \sin 75°$

$\therefore eq = e′e \times 0.966 \quad (e′e = 1500 - 450 = 1050)$

$\qquad = 1050 \times 0.966 = 1014.3$ mm

distance $rt = \sqrt{(te)^2 - (er)^2}$

$rt = \sqrt{(1500)^2 - (1448.9)^2}$

$\qquad = \mathbf{388.2\ mm}$

Ordinate $qr = er - eq = 1448.9 - 1014.3$

$\qquad\qquad = \mathbf{434.6\ mm}$

In triangle *dpt*:

$\angle dpt = 90°$ and $\angle ptd = 60°$

$dp/td = \sin \angle ptd$

$dp = td \times \sin 60°$

$dp = 1500 \times 0.866 = 1299.04$ mm

In triangle *dnd′*:

$dn/d′d = \sin 60°$

$dn = dd′ \times 0.866 = 1050 \times 0.866 = 909.3$ mm

Distance $pt = \sqrt{(td)^2 - (pd)^2}$

$pt = \sqrt{(1500)^2 - (1299.04)^2} = \mathbf{750.00\ mm}$

Ordinate $np = 1299.04 - 909.3 = \mathbf{389.7\ mm}$

Similarly, ordinates *lm*, *jk* and *gh* can be calculated.

lm = 318.1 mm

mt = 1060.8 mm

jk = 225.0 mm

kt = 1299.0 mm

gh = 116.2 mm

ht = 1448.9 mm

Join points *X*, *g*, *j*, *l*, *n*, *q* and *s* to get a smooth curve. These ordinates and the corresponding horizontal distances *ht*, *kt*, *mt*, *pt* and *rt* can be used to set out the other half of the ellipse. Figure 14.11b shows the end result.

Exercise 14.1

The solutions to Exercise 14.1 can be found in Appendix 2.

1. The factory unit *EDCF* in Figure 14.1 measures 36 m by 15 m. Prepare sufficient setting out information to enable the position of this factory unit to be established. Distances from stations *A* and *B* are: *AB* = 116.861 m; *AF* = 74.447 m; *BF* = 45.641 m; *AC* = 109.252 m; *BC* = 11.722 m.

2. Figure 14.12 gives dimensions of a building which is believed to be out of square. Calculate the lengths of sufficient diagonals so that the squareness of the building can be fully checked.

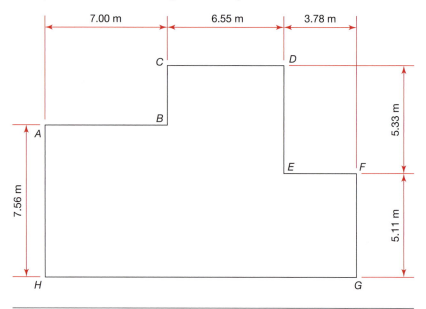

Figure 14.12

3. The plan of a bay window is formed by part of a circular curve of 3.0 m diameter. It is just less than a semicircle, having an inside dimension from chord to quadrant point of 1.4 m. Prepare data in the form of ordinate lengths so that the bricklayers will be able to set out the plan.

4. A circular arch has a span of 2.7 m and a rise of 1.0 m. Find the radius of the arch and the lengths of nine ordinates in order that the arch may be set out.

5. An elliptical arch has a span of 3.6 m and a rise of 1.2 m. Prepare setting-out data for the arch.

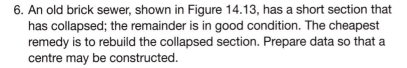

6. An old brick sewer, shown in Figure 14.13, has a short section that has collapsed; the remainder is in good condition. The cheapest remedy is to rebuild the collapsed section. Prepare data so that a centre may be constructed.

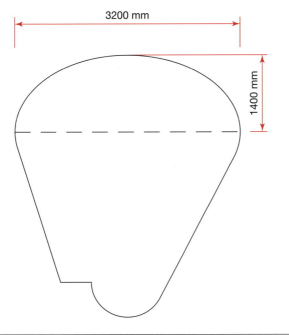

3200 mm

1400 mm

Figure 14.13

Answers to Exercise 14.1

1. See Appendix 2.
2. Diagonal measurements are:

AG = 18.91 m	EH = 14.48 m	CG = 14.69 m
FH = 18.07 m	HB = 10.30 m	BD = 7.16 m
CE = 8.44 m	AC = 7.57 m	EG = 6.36 m
DH = 17.11 m		

3. The five ordinates on the left of the arch centre other than zero and 1.4 m are: 1.379 m, 1.314 m, 1.199 m, 1.018 m and 0.729 m.
4. R = 1.411 m. Splitting half the arch into eight strips, the seven intermediate ordinates are: 0.361 m, 0.572 m, 0.720 m, 0.828 m, 0.906 m, 0.959 m and 0.990 m.
5. Referring to Figure 14.14:

Horizontal distance	Vertical distance
th = 1.739 m	hg = 0.311 m
tk = 1.559 m	kj = 0.600 m
tm = 1.273 m	lm = 0.849 m

tp = 0.900 m np = 1.039 m

tr = 0.466 m rq = 1.159 m

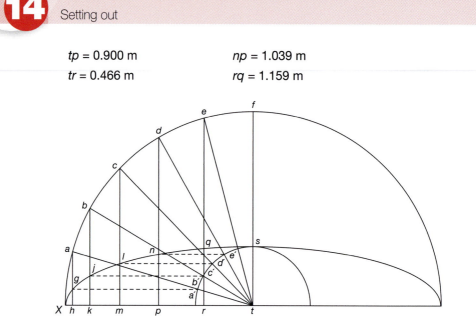

Figure 14.14

6. Horizontal distance Vertical distance

th = 1.545 m hg = 0.362 m

tk = 1.386 m kj = 0.700 m

tm = 1.131 m lm = 0.990 m

tp = 0.800 m np = 1.212 m

tr = 0.414 m rq = 1.352 m

CHAPTER 15

Costing: materials and labour

Learning outcomes:

(a) Calculate the quantities of materials for a range of construction activities

(b) Work out the cost of materials

(c) Calculate the cost of labour for a range of construction activities

15.1 Introduction

One of the important branches of building technology is to prepare the estimates of the cost involved in constructing a building. The overall cost of a project determines its viability and enables the client to arrange capital for financing the construction. The main components of the cost of constructing a building are the cost of materials, labour, plant and contractor's profit.

This chapter gives some examples on how to prepare approximate estimates. The costs of materials and labour are never constant and they also differ from one region to another. For the latest information on the cost of materials, labour and plant, reference should be made to publications which update this information at regular intervals.

15.2 Foundations

Concrete is by far the most commonly used material in the construction of foundations of buildings. For plain concrete, 1:3:6 concrete mix is usually used, but for reinforced concrete 1:2:4 or a stronger mix is used. As described in Chapter 12, the quantities of cement and aggregates for preparing a concrete mix may be determined by considering either their volume or mass. In this section we will consider how to work out the mass of the cement and the aggregates. This gives a better quality of concrete as we can take into account the moisture content of the aggregates if they are not dry.

Consider a 1:3:6 concrete mix. The density of concrete is approximately 2400 kg/m³ or in other words, 1 m³ of concrete has a mass of 2400 kg. To prepare 1 m³ of concrete the quantities of cement and the aggregates are:

$$\text{Cement} = \frac{1}{10} \times 2400 = 240 \text{ kg (the quantity of cement is 1 part out of 10)}$$

$$\text{Fine aggregates} = \frac{3}{10} \times 2400 = 720 \text{ kg}$$

$$\text{Coarse aggregates} = \frac{6}{10} \times 2400 = 1440 \text{ kg}$$

The amount of water depends on where the concrete is to be used. For strip foundations the amount of water could be about 50% of the amount of cement, giving a water–cement ratio of 0.5.

Example 15.1

Calculate the cost of materials and labour required to construct the 1:3:6 concrete strip foundation shown in Figure 15.1, given that:

Cost of materials

Cement = £4.90 per 25 kg bag

Fine aggregates = £36.00 per 850 kg jumbo bag

Coarse aggregates = £36.00 per 850 kg jumbo bag

Labour

Approximate labour hours = 2 per m³ for mixing and placing concrete.

Hourly rate = £15.00

Solution:

Area of foundation = 2[(7.3 + 0.6 + 0.6) × 0.6 + (6.3 × 0.6)]

= 2[(8.5) × 0.6 + (3.78)]

= 17.76 m²

Volume of concrete = 17.76 × thickness

= 17.76 × 0.2 = 3.552 m³

The quantities of materials to prepare 1 m³ of concrete are given in Section 15.2. For preparing 3.552 m³ of concrete the quantities of materials and their costs are:

Material	Mass	
Cement	240 × 3.552 = 852 kg	35 × 25 kg bags
Fine aggregates	720 × 3.552 = 2557 kg	3 × 850 kg jumbo bags
Coarse aggregates	1440 × 3.552 = 5115 kg	6 × 850 kg jumbo bags

Cost of cement = £4.90 × 35 = £171.50

Cost of fine aggregates = £36.00 × 3 = £108.00

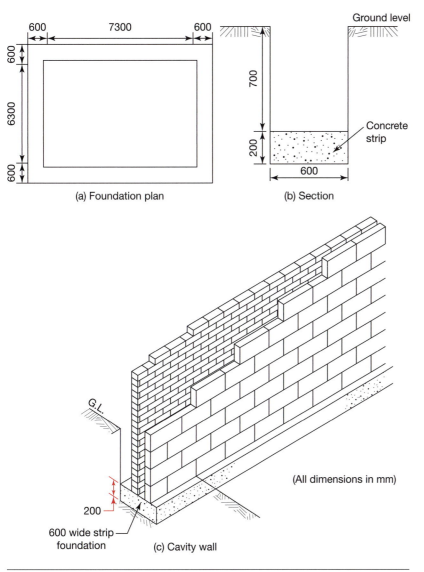

600   7300   600

600

6300

600

(a) Foundation plan

Ground level

700

200

600

Concrete strip

(b) Section

G.L.

200

600 wide strip foundation

(c) Cavity wall

(All dimensions in mm)

Figure 15.1

Cost of coarse aggregates = £36.00 × 6 = £216.00

Total cost of materials = **£495.50**

Labour hours = 2 × 3.552 = 7.104

Labour cost = 7.104 × £15.00 = **£106.56**

Total cost (materials and labour) = £495.50 + £106.56 = **£602.06**

15.3 Cavity walls

The cavity walls in dwelling houses consist of 102.5 mm thick outer leaf of bricks and a 100 mm thick inner leaf of lightweight concrete blocks. The space between the two leaves, called the cavity, is 100 mm wide and provided with mineral wool or other insulation material suitable for the purpose.

Figure 15.2 shows a brick and a block as they are laid in the construction of walls. The dimensions shown are without the mortar joints. With 10 mm thick mortar joints their dimensions are:

Bricks: 225 × 102.5 × 75 mm

Concrete blocks: 450 × 100 × 225 mm

Surface area of one brick = 0.225 × 0.075 m = 0.0169 m²

Number of bricks per m² = $\dfrac{1}{0.0169}$ = 59.26, so 60

Surface area of one concrete block = 0.450 × 0.225 m = 0.10125 m²

Number of concrete blocks per m² = $\dfrac{1}{0.10125}$ = 9.88, so 10

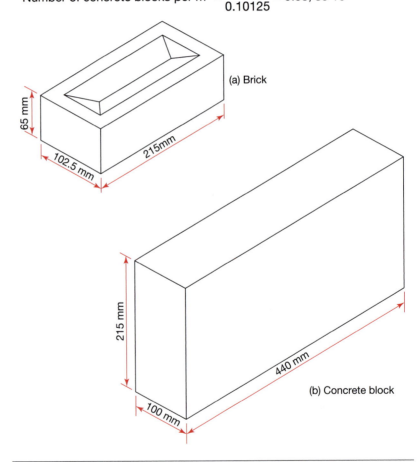

(a) Brick

(b) Concrete block

Figure 15.2

The quantity of mortar required for 1 m² of brickwork is 0.026 m³ and for 1 m² of blockwork is 0.012 m³.

Example 15.2

A traditional cavity wall, 4.0 m × 2.8 m high, has a 2.0 m × 1.2 m high window. Calculate:

(a) The number of bricks and 100 mm thick aerated concrete blocks, and the quantities of cement and sand required to construct the wall. Allow for 5% extra bricks and blocks and 10% extra mortar.

(b) The cost of labour, if the labour rates are: £36.00/m² for brickwork and £9.00/m² for blockwork.

Solution:

Area of the wall = 4.0 × 2.8 − (2.0 × 1.2)

$\quad\quad\quad\quad\quad$ = 11.2 − 2.4 = 8.8 m²

Number of bricks = 8.8 × 60 = 528

Number of bricks with 5% extra allowance = 528 × 1.05 = 554.4, so **555**

Note: *For increasing a quantity by 5%, simply multiply it by 1.05. Similarly for increasing a quantity by 10%, multiply it by 1.10*

The extra allowance of 5% may also be calculated as:

Number of 5% extra bricks = $\dfrac{5}{100}$ × 528 = 26.4, so 27

Number of bricks with 5% extra allowance = 528 + 27 = **555**

Number of blocks = 8.8 × 10 = 88

Number of blocks with 5% extra allowance = 88 × 1.05 = 92.4, so **93**

Mortar required for brickwork = 0.026 × 8.8 = 0.2288 m³

Quantity of mortar with 10% extra allowance = 1.1 × 0.2288 = 0.252 m³

Mortar required for blockwork = 0.012 × 8.8 = 0.1056 m³

Quantity of mortar with 10% extra allowance = 1.1 × 0.1056 = 0.116 m³

Brickwork

Assume the density of 1:3 cement/sand mortar to be 2300 kg/m³

Mass of 0.252 m³ of mortar = 0.252 × 2300 = 579.6 kg

Mass of cement = $\dfrac{1}{4}$ × 579.6 = 144.9 kg

Mass of sand = $\dfrac{3}{4}$ × 579.6 = 434.7 kg

Blockwork

Assume the density of 1:6 cement/sand mortar to be 2300 kg/m³

Mass of 0.116 m³ of mortar = 0.116 × 2300 = 266.8 kg

Mass of cement = $\dfrac{1}{7}$ × 266.8 = 38.11 kg

Mass of sand = $\dfrac{6}{7}$ × 266.8 = 228.69 kg

Total quantity of cement = 144.9 + 38.11 = **183.01 kg**

Total quantity of sand = 434.7 + 228.69 = **663.39 kg**

Cost of labour

Brickwork = £36.00 × 8.8 = **£316.80**

Blockwork = £9.00 × 8.8 = **£79.20**

15.4 Flooring

Softwood boards and chipboard sheets are used as floor-covering materials in dwelling houses. The thickness of these materials depends on the spacing between the floor joists. For a spacing of 400 mm between the joists, 18 mm thick floorboards and chipboard sheets are satisfactory. The coverage provided by tongued and grooved floor boards/chipboard is slightly less than their actual surface area as the tongue of one board fits into the groove of the next one. A floorboard measuring 2400 × 121 × 18 provides coverage of 0.272 m². Figure 15.3 shows the joists and floorboards of an upper floor.

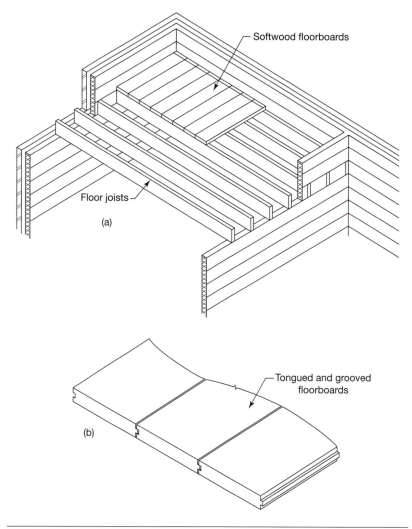

Softwood floorboards

Floor joists

(a)

Tongued and grooved floorboards

(b)

Figure 15.3

Example 15.3

Find the cost of providing and fixing:

(a) Softwood floorboards in a room measuring 4.5 m × 3.9 m.

(b) Chipboard flooring in a room measuring 5.0 m × 4.2 m.

Floorboards measuring 3000 × 121 ×18 mm thick cost £29.50 for a pack of five. One pack covers an area of 1.71 m². Cost of labour is £10.00/m².
  One sheet of chipboard measuring 2400 × 600 × 22 mm thick costs £11.50. One sheet covers an area of 1.44 m². Labour cost is £6.00/m².
  Allow wastage of 10%.

Solution:

(a)

Area of the floor = 4.5 × 3.9 m = 17.55 m²

Number of packs of floorboards = $\dfrac{17.55}{\text{coverage provided by one pack}}$

$$= \dfrac{17.55}{1.71} = 10.26$$

Wastage of 10% = $10.26 \times \dfrac{10}{100} = 1.026$

Total number of packs required = 10.26 + 1.026 = 11.286 or 12

Cost of material = 12 × £29.50 = £354.00

Labour cost = 17.55 × £10.00 = £175.50

Total cost = £354.00 + £175.50 = **£529.50**

(b)

Area of the room = 5.0 × 4.2 = 21.0 m²

Area of one sheet of chipboard = 2.4 × 0.6 = 1.44 m²

Number of sheets required = $\dfrac{21.0}{1.44} = 14.58$

Wastage of 10% = $14.58 \times \dfrac{10}{100} = 1.458$

Total number of chipboard sheets = 14.58 + 1.458 = 16.038 or 16

Cost of material = 16 × £11.50 = £184.00

Labour cost = 21 × £6.00 = £126.00

Total cost = £184.00 + £126.00 = **£310.00**

15.5 Painting

Painting is necessary to give doors, windows, walls, etc. a finish that enhances their appearance and provides protection from dust, dirt and other harmful substances. In this section only the external surfaces of walls are considered. Some of the main factors that affect the amount of paint required are the type of surface, the number of coats and the type of paint.

Example 15.4

The front wall of a house, shown in Figure 15.4, is finished with roughcast rendering and needs two coats of masonry paint. The spreading rate of paint is 3 m² per litre. Find the cost of paint and labour if one five-litre can of masonry paint costs £25.39, and the labour cost is £5.80 per m².

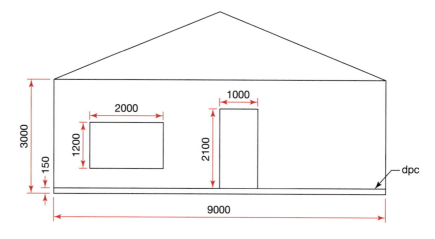

(All dimensions in mm)

Figure 15.4

Solution:

The net area of the wall is calculated by subtracting the area of the door and window from the total wall area:

Area of roughcast finish on the wall = 9.0 × (3.0 − 0.15) − 2.0 × 1.2 − 2.1 × 1.0

= 25.65 − 2.4 − 2.1 = 21.15 m²

$$\text{Volume of paint required} = \frac{\text{area of wall}}{\text{spreading rate of paint}}$$

$$= \frac{21.15}{3} = 7.05 \text{ litres}$$

Two coats of paint require 2 × 7.05 or 14.1 litres of paint

$$\text{Number of five-litre cans} = \frac{14.1}{5} = 2.82, \text{ so } 3$$

Cost of paint = 3 × £25.39 = **£76.17**

Labour cost (two coats) = 21.15 × £5.80 = **£122.67**

Total cost = £76.17 + £122.67 = **£198.84**

Exercise 15.1

The solutions to Exercise 15.1 can be found in Appendix 2.

For the following tasks refer to the relevant example for material and labour costs:

1. Calculate the cost of materials and labour required to construct the 1:3:6 concrete strip foundation shown in Figure 15.5.

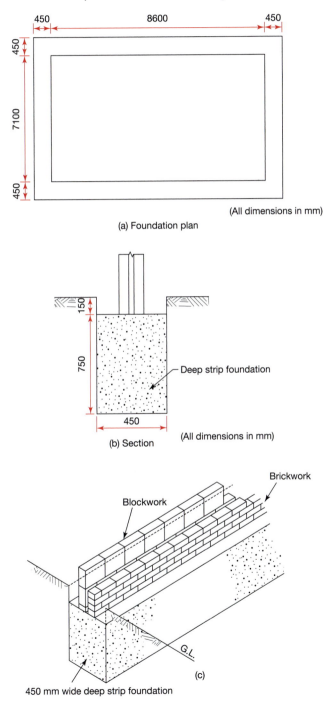

450   8600   450

450

7100

450

(All dimensions in mm)

(a) Foundation plan

150

750

Deep strip foundation

450

(All dimensions in mm)

(b) Section

Brickwork

Blockwork

G.L.

(c)

450 mm wide deep strip foundation

Figure 15.5

2. Calculate the number of bricks and 100 mm thick aerated concrete blocks, and the quantities of cement and sand to construct a 5.5 m × 2.7 m high cavity wall with two 1.80 m × 1.2 m high windows. Allow for 5% extra bricks and blocks and 10% extra mortar.

3. Allowing wastage of 10%, find the cost of providing and fixing:

   (a) softwood floorboards in a room measuring 5.0 m × 4.2 m

   (b) chipboard flooring in a room measuring 4.5 m × 3.9 m.

4. The walls of a house, shown in Figure 15.6, are finished with roughcast rendering and need two coats of masonry paint. The spreading rate of paint is 3 m² per litre. Find the cost of paint and labour if one five-litre can of masonry paint costs £25.39, and the labour cost is £5.80 per m².

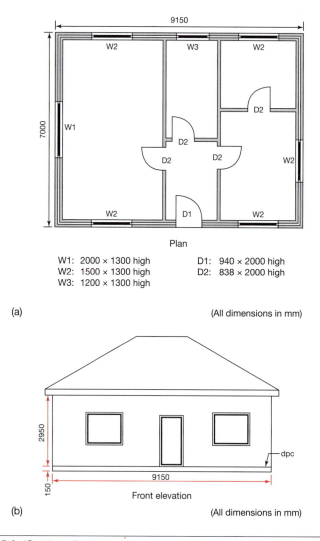

Plan

W1: 2000 × 1300 high   D1: 940 × 2000 high
W2: 1500 × 1300 high   D2: 838 × 2000 high
W3: 1200 × 1300 high

(a)   (All dimensions in mm)

Front elevation

(b)   (All dimensions in mm)

Figure 15.6 *(Continued)*

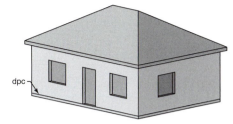

(c)   3D view showing front and side of building

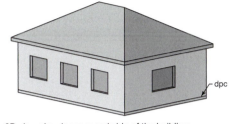

(d)   3D view showing rear and side of the building

Figure 15.6 *(Continued)*

Answers to Exercise 15.1

1. Cost of materials = £1573.20

   Cost of labour = £336.15

2. Number of bricks = 664

   Number of blocks = 111

   Mass of cement = 218.86 kg

   Mass of sand = 793.6 kg

   Labour: brickwork = £379.08; blockwork = £94.77

3. Softwood floorboards: material = £413.00; labour = £210.00

   Chipboard flooring: material = £161.00; labour = £105.30

4. Cost of paint = £279.29; labour cost = £461.07

CHAPTER 16

Statistics

Learning outcomes:

(a) Arrange data into groups and prepare frequency tables

(b) Calculate mean, mode and median of the given data

(c) Present the data/results in the form of a statistical diagram

(d) Prepare histograms, frequency polygons and cumulative frequency curves, and calculate the mode, median and interquartile range

16.1 Introduction

Statistics is a branch of mathematics that involves the collection, preparation, analysis, presentation and interpretation of data. The data may be either primary data or secondary data. Primary data are obtained from people by carrying out surveys via questionnaires and interviews. Primary data are published in many newspapers, journals, magazines and other publications. The collection of primary data can be very time consuming and expensive. The information obtained from opinion polls and market research surveys are typical examples of primary data.

Samples of primary data may be extracted for other purposes. This is known as secondary data.

16.2 Tally charts

The data collected from the questionnaires or publications may be aggregated using a tally chart. For example, in a survey, 25 people shopping in a DIY store were asked about the type of house/building they lived in. Their responses were: semi-detached (semi), detached, terraced, semi, bungalow, flat, detached, flat, terraced, semi, bungalow, flat, semi, terraced, detached, semi, bungalow, flat, terraced, semi, detached, flat, flat, semi, terraced. The five types of accommodation are entered in the tally chart as shown in Table 16.1. To make the totalling easier and less

Table 16.1

Accommodation	Tally	Number or frequency
Flat	ⅢⅠ I	6
Terraced house	ⅢⅠ	5
Semi-detached house	ⅢⅠ II	7
Detached house	IIII	4
Bungalow	III	3

confusing, the numbers are grouped in fives, four vertical bars and the fifth a diagonal/horizontal through the four.

Tally charts may also be used for recording other data, such as traffic counts.

16.3 Tables

Data may also be presented in a tabular form. Table 16.2 shows the variation in daily temperature and humidity in a college building.

Table 16.2

Week number	Day	Temperature (°C)	Relative humidity (%)
1	Monday	20.5	43
	Tuesday	21.6	45
	Wednesday	22.0	44
	Thursday	22.4	46
	Friday	22.1	50
2	Monday	19.0	52
	Tuesday	20.2	51
	Wednesday	21.0	47
	Thursday	21.8	43
	Friday	21.5	45

16.4 Types of data

Depending on the method used, the data collected may be either discrete data or continuous data.

16.4.1 Discrete data

Data collected as integers (whole numbers) is called discrete data. For example, number of employees in construction companies, number of radiators in buildings, number of cars per family, etc.

16.4.2 Continuous data

Continuous data, unlike discrete data, do not increase in jumps, but can have any value between the given limits. For example, height of people, daily temperatures, weight of people, cost of materials, labour costs, etc.

16.4.3 Raw data

In a survey involving the weight (kg) of first-year college students, the following data were obtained:

60.1, 65.5, 63.6, 55.0, 58.8, 61.5, 65.9, 56.6, 55.2, 59.3, 56.5, 64.1, 63.2, 56.0, 58.3, 64.6, 57.4, 63.9, 66.0, 60.7

This is an example of raw data, as it is shown in the manner in which it was collected. It has not been re-arranged to show either ascending/descending values or any other form of arrangement.

16.4.4 Grouped data

The data given in Section 16.4.3 shows that the weight of the students varies between 55.0 kg and 66.0 kg. In order to present and understand the information easily, the data may be arranged in groups. The number of groups will depend on the amount of data. For a small set of data, e.g. 25–50 items, the number of groups may be 5–10. For larger amounts of data the upper limit should be 20. For the data shown in Section 16.4.3, the following groups may be used:

54.0–56.4 kg, 56.5–58.9 kg, 59.0–61.4 kg, 61.5–63.9 kg, 64.0–66.4 kg

A group is called a class and each class is specified by two limits, the lower class limit and the upper class limit. For the first class, 54.0 is the lower and 56.4 the upper limit. Theoretically, this class includes the data between 53.95 and 56.45. Therefore, 53.95 is called the lower class boundary and 56.45 the upper class boundary.

Example 16.1

In a traffic count, the following vehicles passed by the observation point between 0845 and 0846:

Car, heavy goods vehicle (HGV), car, motorbike, HGV, car, car, van, bus, car, bus, car, car, HGV, car, HGV, motorbike, van, bus, car, bus, van, car

Prepare a tally chart.

Solution:

There are five types of vehicles that passed by the observation point. The tally chart is shown in Table 16.3.

Table 16.3

Type of vehicle	Tally	Number or frequency
Motorbike	II	2
Car	ℍℍ ℍℍ	10
Van	III	3
Bus	IIII	4
HGV	IIII	4

Example 16.2

The crushing strengths (unit: N/mm^2) of 50 concrete cubes are given here. Group the data into seven classes and find the frequency of each class.

34	46	40	37	40	35	40	42	34	43	40	45
39	38	46	45	44	34	50	45	35	39	38	35
48	37	42	50	39	46	41	44	41	51	42	47
49	36	47	48	49	50	38	44	44	43	51	34
41	37 N/mm^2										

Solution:

The minimum and the maximum strengths are 34 N/mm^2 and 51 N/mm^2, respectively. The seven classes and their frequencies are shown in Table 16.4.

Table 16.4

Class interval	Tally	Frequency
32–34	IIII	4
35–37	ℍℍ II	7
38–40	ℍℍ ℍℍ	10
41–43	ℍℍ III	8
44–46	ℍℍ ℍℍ	10
47–49	ℍℍ I	6
50–52	ℍℍ	5
		Total = 50

16.5 Averages

After the collection of data, it is often necessary to calculate the average result. In statistics there are three types of average: mean, mode and median.

16.5.1 The mean

The mean, also called the arithmetic mean, is calculated by dividing the sum of all items of the data by the number of items:

$$\text{Mean} = \frac{\text{sum of all items of data}}{\text{number of items in the data}}$$

This can also be written as: $\bar{x} = \frac{\sum x}{n}$ where $\bar{x}$ (read as 'x bar') denotes the mean.

> $\sum x$ means the sum of all items of the data. ($\sum$ (Sigma, a Greek symbol) is used to denote the sum in many mathematical operations) and n denotes the number of items of the data.

In the case of grouped data it becomes necessary to determine the class midpoint. The class midpoint is the average of the lower class boundary and the upper class boundary. For grouped data:

$$\text{Mean, } \bar{x} = \frac{\sum fx}{\sum f}$$

where, x = class midpoint and f = frequency of each class.

16.5.2 The mode

The number that occurs most often in a set of data is called its mode. Sometimes the data may not have a clear mode and sometimes there may be more than one mode. In the case of grouped data, the class having the highest frequency is called the modal class.

16.5.3 The median

If the data are arranged in ascending or descending order, the middle number is called the median. Therefore, the median value of a set of data divides it into two equal halves.

16.5.4 Comparison of mean, mode and median

The mean is the most commonly used average in mathematics, science and engineering. This is the only average that involves all the data, but its value is easily affected if some of the values are extremely high or low as compared to the rest of the data. It can also give an answer that may be impossible, e.g. 2.8 children per family.

The median is not affected by extreme values, but large amounts of data are needed for the median value to be reliable. The median does not involve all the data, but, the answer is an actual value. Its calculation also involves the rearrangement of data in an ascending or descending order.

The mode is easy to find and can be useful when the mean gives meaningless results, as in the example of the number of children per family. The mode of a data set can be an actual value, but sometimes the data may show more than one mode. Like the median, the mode does not involve all the data.

16.6 The range

The range is a measure of the spread or dispersion of data and is defined as the difference between the highest and the lowest values.

Range = highest value – lowest value

Sometimes, two or more sets of data have the same mean, but the data values are very different. For example, two groups of students conducted compression tests on concrete cubes made from the same concrete mix, and obtained the following data:

Group 1: 10, 15, 18, 20, 23, 28 N/mm^2

Group 2: 15, 18, 19, 20, 21, 21 N/mm^2

Both groups obtained the same mean strength of their samples, i.e. 19 N/mm^2, but the actual data were very different. The range of the two sets shows this difference.

Group 1: Range = 28 – 10 = 18 N/mm^2

Group 2: Range = 21 – 15 = 6 N/mm^2

Group 2's observations show more consistency as the range is smaller.

The dispersion of data may also be calculated by determining the interquartile range, which is described in Section 16.8.

Example 16.3

A water absorption test was performed using a sample of 12 bricks. The results, expressed as a percentage, were:

13, 11, 12, 10, 12.5, 12.7, 13, 13.3, 11.2, 13.8, 15 and 13.

Calculate the mean, mode, median and range.

Solution:

(a) Mean:

$$\bar{x} = \frac{\Sigma x}{n}$$

$$= \frac{13+11+12+10+12.5+12.7+13+13.3+11.2+13.8+15+13}{12}$$

$$= \frac{150.5}{12} = \textbf{12.54\%}$$

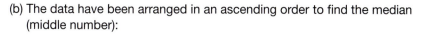

(b) The data have been arranged in an ascending order to find the median (middle number):

10, 11, 11.2, 12, 12.5, 12.7, 13, 13, 13, 13.3, 13.8, 15

There are two middle numbers: 12.7 and 13. The median is the mean of 12.7 and 13, i.e. **12.85%**.

(c) Number 13 occurs three times, all other numbers occur once. Therefore, the mode is **13%**.

(d)

Range = highest number – lowest number

= 15 – 10 = **5**

Example 16.4

In an investigation a group of students was asked to find the compressive strength (N/mm^2) of 40 bricks. The results were:

20	27	21	26	22	25	24	23	32	27
28	27	29	28	27	20	29	28	27	26
30	29	28	27	26	25	30	29	28	27
24	23	35	28	34	29	33	30	32	31

Group the data into six classes and calculate the mean compressive strength.

Solution:

The six classes and their frequencies are shown in Table 16.5.

Table 16.5

Class interval	Class mid-point (x)	Frequency (f)	f × x
19–21	20	3	20 × 3 = 60
22–24	23	5	23 × 5 = 115
25–27	26	12	26 × 12 = 312
28–30	29	14	29 × 14 = 406
31–33	32	4	32 × 4 = 128
34–37	35	2	35 × 2 = 70
		$\Sigma f = 40$	$\Sigma fx = 1091$

Mean, $\bar{x} = \dfrac{\Sigma fx}{\Sigma f} = \dfrac{1091}{40} = $ **27.3 N/mm^2**

16.7 Statistical diagrams

After analysing the data, a variety of statistical diagrams may be used to present the results. The purpose of a diagram is not to display the details within the data, but to show the general pattern. Some commonly used diagrams are pictograms, bar charts, pie charts and line graphs.

16.7.1 Pictograms

Pictograms use pictures to show the information, and hence make the presentation of results more interesting as compared to other diagrams. A lot of care is required in selecting the picture. It should be able to represent all features of the data.

16.7.2 Bar charts

Bar charts consist of data represented in the form of vertical or horizontal bars of equal width. Their heights or lengths vary depending on the quantity they represent. A bar chart may consist of:

- single bars
- multiple bars
- component bars: multiple bars placed directly on top of each other
- Gantt charts: these charts show progress over a period of time
- back-to-back bars.

The first two types are dealt with in Examples 16.5 and 16.6.

16.7.3 Pie charts

A pie chart or a pie diagram is a circle divided into a number of sectors, each sector representing one portion of the data. In order that the sizes of the sectors are in accordance with the quantities they represent, angles at the centre of a pie chart are determined. Several variations of a pie chart can be used, e.g. simple pie, exploded pie, three-dimensional pie, etc. Example 16.5 explains the procedure for producing pie charts.

16.7.4 Line graphs

A detailed discussion on line graphs has already been given in Chapter 8.

Example 16.5

An architects practice employs 4 architect directors, 16 project architects, 20 architect assistants, 10 technicians and 6 administrative staff. Represent the data as: (a) a pictogram, (b) a vertical bar chart, (c) a horizontal bar chart (d) pie charts showing a simple pie and an exploded pie.

Solution:

(a) The pictogram is shown in Figure 16.1.

Category of employees	Number
Architect directors	👤
Project architects	👤 👤 👤 👤
Architect assistants	👤 👤 👤 👤
Technicians	👤 👤
Administrative staff	👤 👤

👤 – Represents one employee

👤 – Represents four employees

👤 – Represents five employees

Figure 16.1

(b) In Figure 16.2 the x-axis represents the staff and the y-axis represents their numbers. The width of a bar is unimportant, but in a chart all bars should be of the same width. The height of the bars is proportional to the number or frequency they represent.

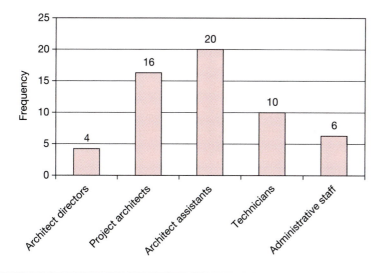

Figure 16.2

(c) In a horizontal bar chart the length of a bar is proportional to the number or frequency it represents, as shown in Figure 16.3.

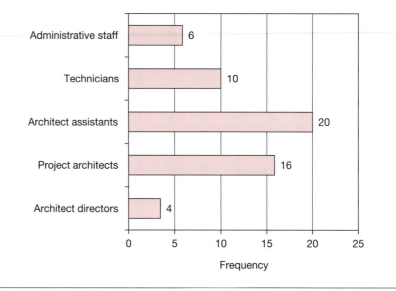

Figure 16.3

(d) There are five types of employee to be shown on the pie chart. The total number of employees is 56, which the circle will represent. The total angle at the centre of a circle is 360°, hence $\left(\dfrac{360}{56}\right)°$ represents one employee. Table 16.6 shows the angles representing the different categories of employees.

Table 16.6

Category of employee	Number of employees	Angle representing each category
Architect directors	4	$= \dfrac{360}{56} \times 4 = 25.7°$
Project architects	16	$= \dfrac{360}{56} \times 16 = 102.9°$
Architect assistants	20	$= \dfrac{360}{56} \times 20 = 128.6°$
Technicians	10	$= \dfrac{360}{56} \times 10 = 64.3°$
Administrative staff	6	$= \dfrac{360}{56} \times 6 = 38.6°$

The pie charts are shown in Figure 16.4. The exploded pie shows the sectors separated from one another.

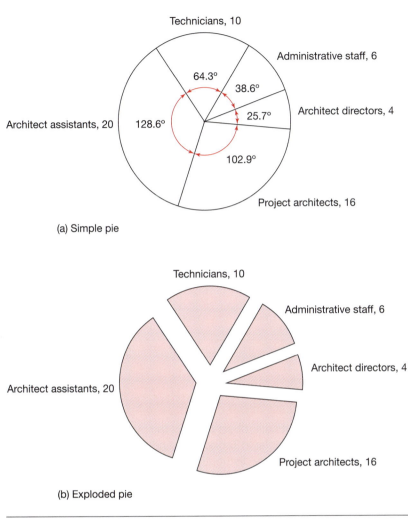

(a) Simple pie

(b) Exploded pie

Figure 16.4

Example 16.6

Table 16.7 shows the typical prices of terraced, semi-detached and detached houses in a British town. These were recorded on 1 July each year, over a period of three consecutive years. Represent the data as a bar chart.

Table 16.7

Year	Price (£)		
	Terraced	Semi-detached	Detached
Year 1	75 000	145 000	200 000
Year 2	90 500	170 000	220 000
Year 3	115 000	190 000	265 000

Solution:

Figure 16.5 shows the prices of the houses as a bar chart with multiple bars.

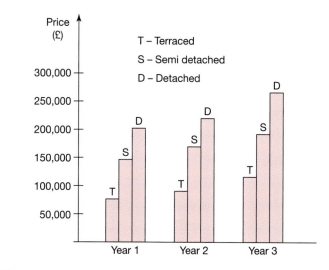

Figure 16.5

16.8 Frequency distributions

The frequency of a particular observation is the number of times it occurs in the data, as shown earlier in Tables 16.1 and 16.3. There are many ways in which the frequency distribution of the given data may be shown, some of which are frequency tables, histograms, frequency polygons and cumulative frequency polygons. The frequency tables are the same as tally charts, as shown in Tables 16.3 and 16.4.

16.8.1 Histograms

A histogram is like a bar chart but with connected bars or rectangles. In a bar chart a scale is only used for the *y*-axis, but in a histogram scales are used for both axes. This means that a histogram can be used to display continuous data. The area of a bar is equal to the class interval multiplied by frequency. Unlike a bar chart, the bars of a histogram can be of different widths. Usually the class intervals in a frequency distribution are of equal widths, in which case the heights of the bars represent the frequencies, as shown in Figure 16.6. A histogram can also be used to determine the mode of grouped data.

16.8.2 Frequency polygons

A frequency polygon is another method of representing a frequency distribution. It is drawn by plotting the frequencies at the midpoints of the groups, and joining the points by straight lines (see Figure 16.7a). Alternatively, a frequency polygon can be constructed by joining the mid points of the histogram bars by straight lines. As a polygon should be closed, extra class intervals are added and the straight lines extended to meet the x-axis, as shown in Figure 16.7b.

Example 16.7

Thirty-two samples of timber were tested to determine their moisture content, the results of which are given in Table 16.8. Display the results as (a) a histogram and (b) a frequency polygon. Use the histogram to calculate the mode value.

Table 16.8

Moisture content (%)	8–10	11–13	14–16	17–19	20–22
Frequency	4	7	11	8	2

Solution:

(a) In a histogram the frequency is shown on the y-axis and the other quantity on the x-axis. In this example the moisture content is shown on the x-axis and the frequency on the y-axis. The next step is to select suitable scales and mark them on the x-axis and y-axis.

As the group sizes (or class intervals) are the same, bars of uniform width will be used. To show the first bar two vertical lines are drawn, one at the lower class boundary (7.5%) and the other at the upper class boundary (10.5%) of the first class. The height of the bar should be equal to the frequency of the group. Therefore, a horizontal line is drawn from the point that represents a frequency of 4. This produces the first bar; similarly, the rest of the diagram is drawn as shown in Figure 16.6.

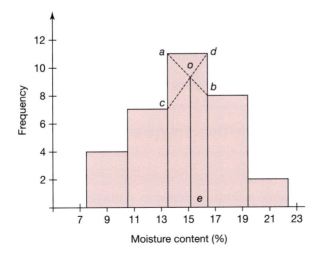

Figure 16.6

(b) A frequency polygon may be drawn as explained above. The first step is to determine the class midpoints and then to plot them against relevant frequencies. The class midpoints in this example are 9, 12, 15, 18 and 21. The points are joined by straight lines, as shown in Figure 16.7a. As a polygon should be closed, extra class intervals are added and the straight lines extended to meet the x-axis, as shown in Figure 16.7b.

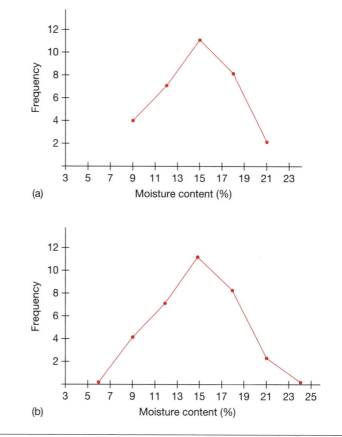

Figure 16.7

To determine the mode value, draw diagonal lines *ab* and *cd* in the tallest bar. Draw a vertical line from the point of intersection, *o*, to meet the x-axis at *e*. Point *e* gives the mode value, which in this case is **15.2%**.

16.8.3 Cumulative frequency distribution

A cumulative frequency curve may also be used to represent a frequency distribution. To draw a cumulative frequency curve, the data are grouped into an appropriate number of classes as in the case of frequency polygons. Cumulative frequency is then determined for each class as a number less than the upper class boundary. The cumulative frequencies are plotted against corresponding upper class boundaries and the points joined by a smooth curve. The curve is called a cumulative frequency curve or ogive.

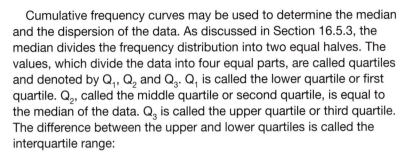

Cumulative frequency curves may be used to determine the median and the dispersion of the data. As discussed in Section 16.5.3, the median divides the frequency distribution into two equal halves. The values, which divide the data into four equal parts, are called quartiles and denoted by Q_1, Q_2 and Q_3. Q_1 is called the lower quartile or first quartile. Q_2, called the middle quartile or second quartile, is equal to the median of the data. Q_3 is called the upper quartile or third quartile. The difference between the upper and lower quartiles is called the interquartile range:

Interquartile range = $Q_3 - Q_1$

Interquartile range is one of the measures used to determine the dispersion of the data. The main advantage of using the interquartile range is that only the middle half of the data is used and thus this measure is considered to be more representative than the range

Example 16.8

Use the data given in Example 16.7 to draw a cumulative frequency curve, and find:

(a) the median moisture content

(b) the interquartile range

(c) the number of samples with moisture content less than 14%.

(d) the number of samples with moisture content more than 18%.

Solution:

Before the data are plotted we need to find the cumulative frequency for each class (or group). This is shown in Table 16.9. The first two columns show the same data as displayed in table 16.8.

Table 16.9

Class interval (moisture content)	Frequency	Moisture content – less than	Cumulative frequency
		7.5	0
8–10	4	10.5	4
11–13	7	13.5	4 + 7 = 11
14–16	11	16.5	11 + 11 = 22
17–19	8	19.5	22 + 8 = 30
20–22	2	22.5	30 + 2 = 32

The cumulative frequencies are calculated as shown and plotted against the upper class boundaries. A smooth curve is drawn through the points to get the cumulative frequency curve. The cumulative frequency is divided into four parts, as shown in Figure 16.8.

$Q_1 = 8$, $Q_2 = 16$ and $Q_3 = 24$.

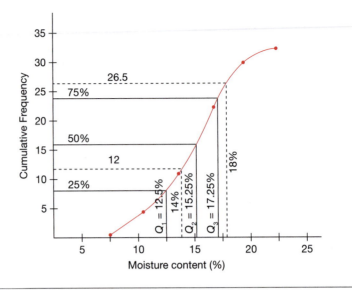

Figure 16.8

(a) Median moisture content = **15.25%** (Q2)

(b) Interquartile range $= Q_3 - Q_1$
$$= 17.25 - 12.5 = \textbf{4.75}$$

(c) The cumulative frequency corresponding to 14% moisture content is 12. Therefore, the moisture content of 12 samples is below **14%**.

(d) The cumulative frequency corresponding to 18% moisture content is 26.5. Therefore, the number of samples with moisture content more than 18% is 32 − 26.5 = 5.5 or **6**.

Exercise 16.1

The solutions to Exercise 16.1 can be found in Appendix 2.

1. Twelve samples of sand were tested to determine their silt content. The amounts of silt, expressed as a percentage, were: 6.0, 8.5, 7.1, 6.3, 5.5, 7.8, 8.2, 6.5, 6.0, 7.7, 8.0 and 7.2. Calculate the mean, mode and median silt contents.

2. The moisture contents (%) of 15 samples of timber were determined in a laboratory and are shown below. Calculate the mean, mode, median and range: 11.5, 14, 11, 12, 13.5, 10, 12.5, 12.8, 13, 13.3, 11.4, 13.6, 15, 10.5 and 9.8.

3. The crushing strengths (unit: N/mm^2) of 45 concrete cubes are given here. Group the data into seven classes and find:

   (a) the frequency of each class

   (b) the mean crushing strength.

34	40	45	39	35	38	46	45	44
34	50	45	35	39	38	35	37	39
43	48	37	42	50	39	46	41	44
41	51	42	47	49	36	47	48	49
50	38	44	44	43	51	34	41	37

4. The number of personnel in S & R Consulting Engineers is given in the table:

Directors	Associate directors	Civil engineers	Technician engineers	Technicians	Administrative staff
3	4	3	3	4	4

Represent this information in the form of

(a) a vertical bar chart

(b) a horizontal bar chart

(c) a pie chart.

5. The table below shows the typical prices of terraced, semi-detached and detached houses in a British town. These were recorded on 1 August each year over a period of three consecutive years. Represent the data as a bar chart.

Year	Price (£)		
	Terraced	Semi-detached	Detached
Year 1	80 000	150 000	220 000
Year 2	92 500	185 000	235 000
Year 3	120 000	200 000	270 000

6. A multinational construction company has recently published the annual data on the number of accidents that occurred at its construction sites. The following extract shows the number of accidents that occurred due to trenches and excavations:

Depth of trench/ excavation	0.1–1.5 m	1.6–3.0 m	3.1–4.5 m	4.6–6.0 m	6.1 m and over
Number of accidents	32	23	22	9	4

Represent the data as:

(a) a pie chart

(b) a horizontal bar chart.

7. A group of construction students were asked to find the level of illuminance (unit: lux) in different rooms of a building. Their observations were:

| 370 | 400 | 500 | 450 | 425 | 500 | 400 |
| 550 | 450 | 350 | 520 | 450 | 460 | 350 |

410	500	410	330	460	400	380
430	470	370	390	420	360	460
380	450	440	360	470	490	510
460	530	440	480	440		

(a) Group the data into 6–8 classes and prepare a frequency chart.

(b) Display the illuminance levels as a histogram and a frequency polygon.

(c) Determine the mean and mode of the data.

8. The table below shows the compressive strength of 35 samples of concrete (1:2:4 concrete):

Compressive strength (N/mm²)	19–22	23–26	27–30	31–34	35–38
Frequency	3	7	15	6	4

(a) Display the compressive strength as a histogram and a cumulative frequency curve.

(b) Determine the mean, mode and median of the data.

9. The following table shows the compressive strength of class-B engineering bricks:

Compressive strength (N/mm²)	34–37	38–41	42–45	46–49	50–53	54–57	58–61
Frequency	2	8	15	20	13	10	2

(a) Prepare a cumulative frequency chart and plot the data to produce a cumulative frequency curve.

(b) Find the median compressive strength and the interquartile range.

Answers to Exercise 16.1

1. Mean = 7.07%; median = 7.15%; mode = 6.0%

2. Mean = 12.26%; there is no mode; median = 12.5%; range = 5.2%

3. (a) See Appendix 2

   (b) Mean = 42.33 N/mm²

4. (a) See bar chart in Appendix 2

   (b) See bar chart in Appendix 2

   (c) See pie chart in Appendix 2

5. See Appendix 2

6. (a) See pie chart in Appendix 2

   (b) See bar charts in Appendix 2

7. (a) See Appendix 2

(b) See Appendix 2

(c) Mean = 440.5 lux; Mode = 456 lux

8. (a) See Appendix 2

(b) Mean = 28.6 N/mm^2; Mode = 28.5 N/mm^2; Median = 28.5 N/mm^2

9. (a) See ogive in Appendix 2

(b) Median = 47.5 N/mm^2; Interquartile range = 7.5

CHAPTER 17

Areas and volumes (2)

Learning outcomes:

(a) Perform calculations to determine the surface area of cones and pyramids

(b) Perform calculations to determine the surface area and volume of a frustum of a pyramid

(c) Perform calculations to determine the surface area and volume of a frustum of a cone

17.1 Introduction

In this chapter we shall calculate the surface area of cones, pyramids and frusta of cones and pyramids. We shall define the frustum of a pyramid/cone and use the standard formulae to determine their surface areas and the volume enclosed. There are many practical problems that can be solved by using the methods explained in this chapter. As the solution to these problems requires the use of complex formulae, this chapter will be more relevant to the students on level 3 and level 4 courses.

17.2 Surface area of a pyramid

The base of a pyramid may be a triangle, rectangle, square or any other polygon, but the lateral faces in each case are triangles. As the lateral faces slope at a certain angle, we must determine their true length before we can use the formula:

$$\text{Area of a triangle} = \frac{\text{base} \times \text{height}}{2}.$$

If the vertex is vertically above the centre of the base, the pyramid is known as a right pyramid. A pyramid is also known as a regular pyramid if its base is a regular figure.

Example 17.1

The roof of a building is in the form of a pyramid. If the base measures 30 m × 30 m, and the perpendicular height is 7.0 m, calculate the area of the roof covering.

Solution:

Figure 17.1 shows the elevation and sectional view of the roof. Use Pythagoras' Theorem to find the true length of the sloping face.

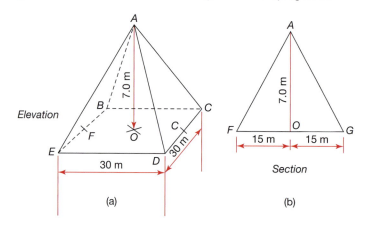

(a)

(b)

Figure 17.1

$$(AF)^2 = (AO)^2 + (OF)^2$$
$$= (7)^2 + (15)^2 = 274$$
$$AF = \sqrt{274} = 16.55 \text{ m}$$

$$\text{Area of face } ABE = \frac{\text{base} \times \text{height}}{2}$$

$$= \frac{30 \times 16.55}{2} = 248.25 \text{ m}^2$$

Due to the symmetry of the (pyramid) roof, all lateral faces are equal.

Total area of roof covering = 4 × 248.25 = **993 m²**

17.2.1 Frustum of a pyramid

Imagine a pyramid being cut by a plane parallel to the base. The portion of the pyramid between the cutting plane *JKLM* and base *EFGH* is known as a frustum. In this case the lateral faces are not triangles but trapezia, as shown in Figure 17.2.

The volume (*V*) of a frustum of a pyramid is given by:

$$\text{Volume} = \frac{1}{3} h(A + \sqrt{AB} + B)$$

where *A* and *B* are the areas of the base and the top (see Figure 17.2) and *h* is the vertical height of the frustum.

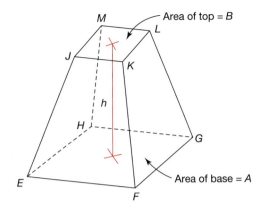

Area of top = B

Area of base = A

Figure 17.2

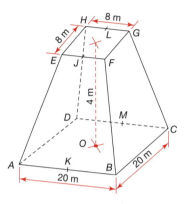

Figure 17.3

Example 17.2

The roof of an old building is shaped like a frustum of a pyramid (Figure 17.3). Find the surface area of the roof and the volume enclosed.

Solution:

Figure 17.4 shows a vertical section through the middle of the frustum.

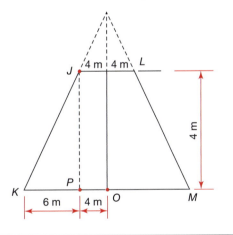

Figure 17.4

Surface area:

Length of the lateral side $(JK) = \sqrt{(JP)^2 + (PK)^2}$

$$= \sqrt{(4)^2 + (6)^2}$$

$$= 7.21 \text{ m}$$

Area of side $EFBA = \dfrac{1}{2}(8 + 20) \times 7.21 = 100.94 \text{ m}^2$

Total surface area = area of lateral sides + area of the top

$$= (4 \times 100.94) + (8 \times 8)$$

$$= \mathbf{467.76 \text{ m}^2}$$

Volume enclosed by the roof:

$$\text{Volume} = \frac{1}{3}h(A + \sqrt{AB} + B)$$

Vertical height of the roof = 4 m

Area of the top (B) = 8 × 8 = 64 m²

Area of the base (A) = 20 × 20 = 400 m²

$$\text{Volume} = \frac{1}{3}4(400 + \sqrt{400\times64} + 64)$$

$$= \frac{1}{3}4(400 + 160 + 64)$$

$$= \frac{1}{3}4(624) = \textbf{832 m}^3$$

17.3 Surface area of a cone

If a cone (Figure 17.5a) is made of paper and the paper is cut along the joint, the shape of the paper will be like the sector of a circle, as shown in Figure 17.5b.

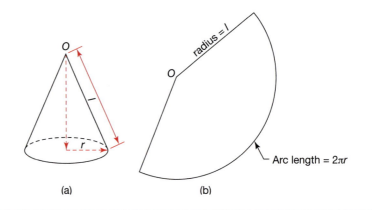

(a)　　　　　(b)

Figure 17.5

Lateral surface area = $\pi r l$

where, r = radius of the base, l = slant height.

Example 17.3

A historical building has circular towers at each corner. The roof of each tower is in the form of a regular cone. If the diameter of one of the towers is 10 m and the height of its conical roof is 4.0 m, find the surface area of the roof.

Solution:

Radius of the base = Diameter ÷ 2

= 10 ÷ 2 = 5 m

Height of the roof = 4.0 m

Slant height = $\sqrt{4^2 + 5^2}$ = 6.403 m

Surface area of the roof = $\pi r l$

$\qquad$ = $\pi \times 5 \times 6.403$ = **100.58 m²**

17.3.1 Frustum of a cone

The area of the lateral surface of a conical frustum (Figure 17.6) is given by:

Area = $\pi l(r + R)$

Total surface area = area of the top + $\pi l(r + R)$ + area of the base

$\qquad$ = $\pi r^2 + \pi l(r + R) + \pi R^2$

Volume = $\dfrac{1}{3}\pi h(r^2 + rR + R^2)$

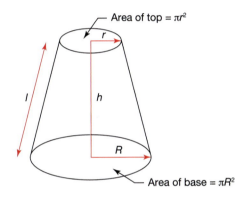

Figure 17.6

Example 17.4

The roof of a part of a historical building is shaped like a frustum of a cone, as shown in Figure 17.7. Find the surface area of the roof and the volume of the space enclosed.

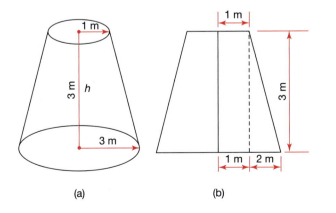

(a) $\qquad$ (b)

Figure 17.7

Solution:

Surface area of the roof:

Slant height $(l) = \sqrt{3^2 + 2^2} = 3.606$ m

Total surface area of the roof $= \pi r^2 + \pi l(r + R)$

$$= \pi(1)^2 + \pi \times 3.606 \, (1 + 3)$$

$$= 3.142 + 45.314$$

$$= \textbf{48.456 m}^2$$

Volume enclosed by the roof:

$$\text{Volume} = \frac{1}{3} \pi h(r^2 + rR + R^2)$$

$$= \frac{1}{3} \pi \times 3(1^2 + (1 \times 3) + 3^2)$$

$$= \frac{1}{3} \pi \times 3(1 + 3 + 9)$$

$$= \frac{1}{3} \pi \times 3(13) = \textbf{40.84 m}^3$$

Exercise 17.1

The solutions to Exercise 17.1 can be found in Appendix 2.

1. The roof of a building is in the form of a pyramid. If the base measures 35 m × 35 m, and the perpendicular height is 7.5 m, calculate the surface area of the roof.

2. A multi-storey building is shaped like a cone. The height of the building is 60 m and the diameter of its base 50 m. Calculate its surface area.

3. Find the area of external cladding for the building in question 2 if the doors and windows have a combined area of 500 m².

4. The roof of a building is shaped like the frustum of a pyramid. The dimensions of the base and the top are 45 m × 45 m and 5 m × 5 m, respectively. If the perpendicular height of the roof is 6 m, calculate:

   (a) the surface area of the roof

   (b) the volume enclosed by the roof.

5. During the process of making ready-mixed concrete the dry mixture is placed in a container which is in the form of an inverted frustum of a right cone 4.5 m high while the top and bottom diameters are 4 m and 2 m, respectively. The base is sealed to another vessel in the form of a shallow inverted right cone 2 m diameter but only 0.5 m high. What is the combined volume of the two vessels?

6. What will be the combined curved surface area of the two containers in question 5?

7. Two water filters are designed to fit into a 150 mm diameter pipe. One consists of a right circular cone 150 mm diameter at its base and 350 mm long. The other consists of a frustum of the same cone with

the smaller end sealed. The length of the frustum is 200 mm. Its whole surface area acts as a filter. Find:

(a) the diameter of the smaller end of the frustum

(b) which filter has the greater surface area and by how much.

Answers to Exercise 17.1

1. 1332.76 m^2

2. 5105.09 m^2

3. 4605.09 m^2

4. (a) 2113.06 m^2; (b) 4550 m^3

5. 33.51 m^3

6. 46.96 m^2

7. (a) 64.29 mm; (b) The cone has a larger surface area by 12 245.11 mm^2

CHAPTER 18

Areas and volumes (3)

Learning outcomes:

(a) Calculate the area of irregular shapes using the mid-ordinate rule, trapezoidal rule and Simpson's rule

(b) Use the trapezoidal rule and Simpson's rule to determine the volume of irregular solids and their application to practical problems

(c) Apply the prismoidal rule to determine the volume of trenches, cuttings and embankments

18.1 Introduction

There are many situations in which we are faced with the determination of the area of plane figures which are bounded by straight lines, and curved lines that do not follow any regular geometrical pattern. The approximate area of an irregular shape can be determined easily by using a range of methods such as graphical technique, planimeter, mid-ordinate rule, trapezoidal rule and Simpson's rule. In this chapter the last three are explained with examples, and later their use is extended to determine the volume of irregular objects.

18.2 Mid-ordinate rule

The irregular figure is divided into a number of strips of equal width by vertical lines, called ordinates. Each strip is considered as approximating to a rectangle of length equal to its mid-ordinate, as shown in Figure 18.1. The mid-ordinate of each strip can be determined by calculating the average of its ordinates. The area of a figure is finally calculated by adding the areas of all the strips.

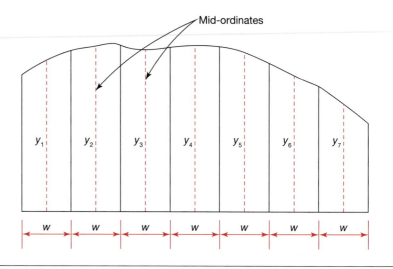

Figure 18.1

Area = $wy_1 + wy_2 + wy_3 + wy_4 + \ldots + wy_7$

$\quad\quad = w\,(y_1 + y_2 + y_3 + y_4 + \ldots + y_7)$

Area = width of strip × sum of mid-ordinates

18.3 Trapezoidal rule

The irregular figure is divided into a number of strips of equal width w (Figure 18.2). Each strip is assumed to be a trapezium. Consider the first strip $ACDB$; its parallel sides are y_1 and y_2 and the distance between them is w.

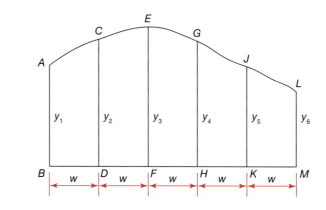

Figure 18.2

Area of strip $ACDB = \dfrac{1}{2}\,(y_1 + y_2)\,w$

Area of strip $CEFD = \dfrac{1}{2}\,(y_2 + y_3)\,w$

Similarly, the area of other strips can be calculated. Finally, all the areas are added to find the area of figure *ALMB*.

Area of figure *ALMB*

$$= \frac{1}{2}(y_1 + y_2)w + \frac{1}{2}(y_2 + y_3)w + \ldots + \frac{1}{2}(y_5 + y_6)w$$

$$= w\left(\frac{y_1 + y_2 + y_2 + y_3 + y_3 + y_4 + y_4 + y_5 + y_5 + y_6}{2}\right)$$

$$= w\left(\frac{y_1 + y_6}{2} + \frac{2y_2}{2} + \frac{2y_3}{2} + \frac{2y_4}{2} + \frac{2y_5}{2}\right)$$

$$= w\left(\frac{y_1 + y_6}{2} + y_2 + y_3 + y_4 + y_5\right)$$

= width of strip $\times$ [$\frac{1}{2}$ (sum of the first and last ordinates) + (sum of the remaining ordinates)]

18.4 Simpson's rule

This method is more complicated than the previous methods but gives a more accurate answer. The figure is divided into an even number of vertical strips of equal width (*w*), giving an odd number of ordinates (Simpson's rule will not work with an even number of ordinates).

Area = $\frac{1}{3}$ (width of strip) [(first + last ordinate) + 4 (sum of the even ordinates) + 2 (sum of the remaining odd ordinates)]

Example 18.1

Find the area of the irregular shape shown in Figure 18.3 using the mid-ordinate rule, the trapezoidal rule and Simpson's rule. Compare the results obtained if the exact answer is 25.33 cm².

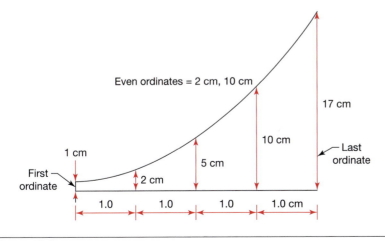

Figure 18.3

Solution:

Mid-ordinate rule:

Mid-ordinate of the first strip $= \dfrac{1 + 2}{2} = 1.5$

Mid-ordinate of the second strip $= \dfrac{2 + 5}{2} = 3.5$

Mid-ordinates of the third and fourth strip are 7.5 and 13.5, respectively.

Area = width of strip × sum of mid-ordinates

$= 1.0 \times (1.5 + 3.5 + 7.5 + 13.5) =$ **26 m²**

Trapezoidal rule:

Area = width of strip $\times \left[\dfrac{1}{2} \text{ (sum of first and last ordinates) + (sum of the remaining ordinates}\right]$

$= 1.0 \times \left[\dfrac{1}{2}(1 + 17) + (2 + 5 + 10)\right]$

$= 1.0 \times [9 + 17] =$ **26 m²**

Simpson's rule:

Area $= \dfrac{1}{3} \times$ (width of strip) [(first + last ordinate) + 4 (sum of the even ordinates) + 2 (sum of the remaining odd ordinates)]

$= \dfrac{1}{3}(1.0) \, [(1 + 17) + 4 \, (2 + 10) + 2 \, (5)]$

$= \dfrac{1}{3}(1.0) \, [(18) + 48 + 10]$

$= \dfrac{1}{3}(1.0) \, [76] =$ **25.33 m²**

The answer produced by the mid-ordinate rule as well as the trapezoidal rule is 26 m², which is higher than the exact area. Simpson's rule, considered to be more accurate than the other two methods, has produced an accurate answer in this example.

Example 18.2

Calculate the area of the plot of land shown in Figure 18.4

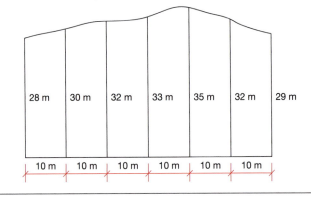

Figure 18.4

Solution:

Mid-ordinate rule:

Area = width of strip × sum of mid-ordinates
= 10 (29 + 31 + 32.5 + 34 + 33.5 + 30.5)
= 10 (190.5) = **1905 m²**

Trapezoidal rule:

Area = width of strip × $\left[\dfrac{1}{2}\right.$ (sum of the first and last ordinates) + (sum of

the remaining ordinates)$\left.\right]$

$= 10 \times \left[\dfrac{1}{2}(28 + 29) + (30 + 32 + 33 + 35 + 32)\right]$

= 10 [190.5] = **1905 m²**

Simpson's rule:

Area $= \dfrac{1}{3} \times$ (width of strip) [(first + last ordinate) + 4 (sum of the even

ordinates) + 2 (sum of the remaining odd ordinates)]

$= \dfrac{1}{3}(10) [(28 + 29) + 4 (30 + 33 + 32) + 2 (32 + 35)]$

$= \dfrac{1}{3}(10) [(57) + 380 + 134]$

$= \dfrac{1}{3}(10) [571] =$ **1903.33 m²**

18.5 Volume of irregular solids

The volume of irregular solids may be determined by any of the three methods explained in the previous section, i.e. the mid-ordinate rule, the trapezoidal rule and Simpson's rule. The only difference is that rather than the ordinates, the cross-sectional areas are used in the volume calculations.

Trapezoidal rule:

Volume = Width of strip × $\left[\dfrac{1}{2}\right.$ (first area + last area) + (sum of the

remaining areas)$\left.\right]$

Simpson's rule:

Volume $= \dfrac{1}{3}$ (width of strip) × [(first area + last area) + 4 (sum of the

even areas) + 2 (sum of the remaining odd areas)]

Example 18.3

The cross-sectional areas of a trench at 10 m intervals are shown in Figure 18.5. Use the trapezoidal rule and Simpson's rule to calculate the volume of earth excavated.

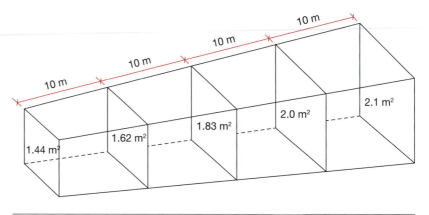

Figure 18.5

Solution:

Trapezoidal rule:

$A_1 = 1.44$ m², $A_2 = 1.62$ m², $A_3 = 1.83$ m², $A_4 = 2.0$ m², $A_5 = 2.1$ m²

$$\text{Volume} = 10 \times \left[\frac{1}{2}(1.44 + 2.1) + (1.62 + 1.83 + 2.0) \right]$$

$$\text{Volume} = 10 \times [(1.77) + (1.62 + 1.83 + 2.0)]$$

$$= 10\,[1.77 + 5.45] = \mathbf{72.2\ m^3}$$

Simpson's rule:

$$\text{Volume} = \frac{1}{3}(\text{width of strip})\,[(\text{first area + last area}) + 4\,(\text{sum of the even}$$
$$\text{areas}) + 2\,(\text{sum of the remaining odd areas})]$$

$$= \frac{1}{3}(10)\,[(1.44 + 2.1) + 4\,(1.62 + 2.0) + 2\,(1.83)]$$

$$= \frac{1}{3}(10)\,[(3.54) + 4\,(3.62) + 2\,(1.83)]$$

$$= \frac{1}{3}(10)\,[3.54 + 14.48 + 3.66]$$

$$= \frac{1}{3}(10)\,[21.68] = \mathbf{72.267\ m^3}$$

18.6 Prismoidal rule

The prismoidal rule is basically Simpson's rule for two strips, and may be used to determine the volume of prisms, pyramids, cones and frusta of cones and pyramids.

$$\text{Volume, by prismoidal rule} = \frac{1}{3}(\text{width of strip}) \times [(\text{area of one end} + 4$$
$$(\text{area of the middle section}) + \text{area of the other end}]$$

$$= \frac{1}{3}(\text{width of strip})\,[A_1 + 4\,(A_2) + A_3]$$

where A_1 = area of one end of the object; A_2 = area of the mid-section; A_3 = area of the other end of the object

If h is the height or length of the solid (see Figure 18.6), the height/length of one strip is $\dfrac{h}{2}$.

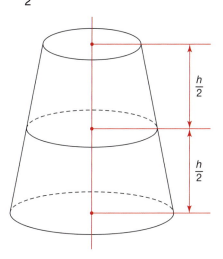

Figure 18.6

$$\text{Volume} = \frac{1}{3}\left(\frac{h}{2}\right)[A_1 + 4\,(A_2) + A_3]$$

$$= \frac{h}{6}[A_1 + 4\,(A_2) + A_3]$$

Example 18.4

An excavation for a basement measures 50 m × 40 m at the top and 44 m × 34 m at the base. If the depth of the excavation is 3.0 m, use the prismoidal rule to calculate the volume of the soil to be excavated.

Solution:

The dimensions at the mid-section of the excavation are:

$$\text{Length} = \frac{50+44}{2} = 47 \text{ m; width} = \frac{40+34}{2} = 37 \text{ m}$$

Depth of the excavation, h = 3.0 m

Area of top (A_1) = 50 × 40 = 2000 m^2

Area of mid-section (A_2) = 47 × 37 = 1739 m^2

Area of base (A_3) = 44 × 34 = 1496 m^2

$$\text{Volume} = \frac{h}{6}[A_1 + 4\,(A_2) + A_3]$$

$$= \frac{3}{6}[2000 + 4\,(1739) + 1496]$$

$$= \frac{3}{6}[10\,452] = \mathbf{5226 \text{ m}^3}$$

Exercise 18.1

The solutions to Exercise 18.1 can be found in Appendix 2.

1. Find the area of the irregular shape shown in Figure 18.7, using the mid-ordinate rule and the trapezoidal rule. Compare the results obtained if the exact answer is 25.833 cm².

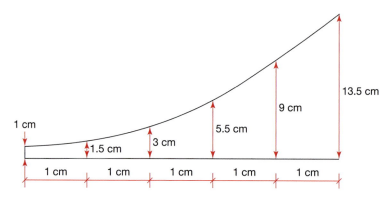

Figure 18.7

2. Find the area of the building plot shown in Figure 18.8 using any two methods and compare the results obtained.

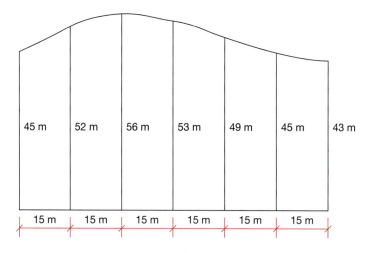

Figure 18.8

3. Find the area of the irregular shape shown in Figure 18.9 by using Simpson's rule.

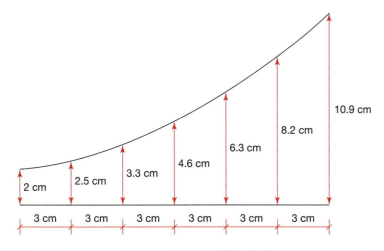

Figure 18.9

4. Figure 18.10 shows the longitudinal section of a trench to be excavated in undulating ground. The depths of the drain below the ground surface are shown at intervals of 12 m. Calculate:

   (a) the area of the section, by the trapezoidal rule

   (b) the volume of the earth to be excavated – width of the trench = 1.2 m.

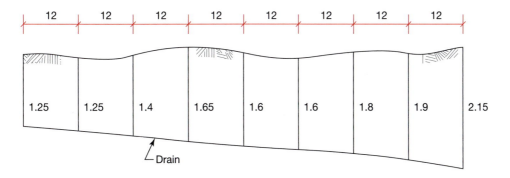

(All measurements in metres)

Figure 18.10

5. Calculate the area of the building plot shown in Figure 18.11.

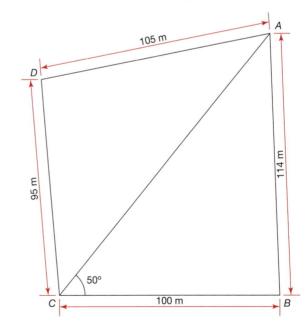

(All measurements in metres)

Figure 18.11

6. Draw a scale diagram of the building plot shown in Figure 18.12 and divide it into a number of strips. Find the length of the ordinates and hence calculate the area of the building plot.

Figure 18.12

7. A 20 m wide excavation is to be made in undulating ground for the construction of a road. The depths of the excavation at intervals of 20 m are shown in Figure 18.13. Use Simpson's rule to calculate the volume of the soil to be excavated. Assume the sides of the excavation to be vertical.

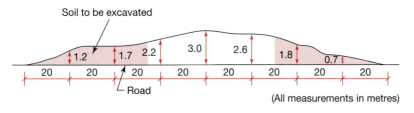

Figure 18.13

8. Figure 18.14 shows the cross-sectional area of sections taken at intervals of 10 m along a trench. Use Simpson's rule to determine the volume of the soil to be excavated.

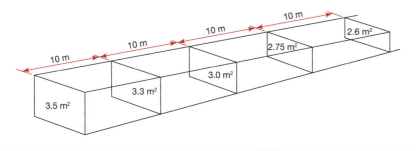

Figure 18.14

9. The basement of a building requires an excavation with battered sides, as shown in Figure 18.15. Calculate the volume of earth to be excavated if the depth of the trench is 3.3 m.

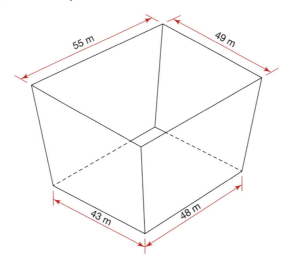

Figure 18.15

10. The plan of an excavation is shown in Figure 18.16, with the depth at the corners circled. What is the volume of the soil to be excavated?

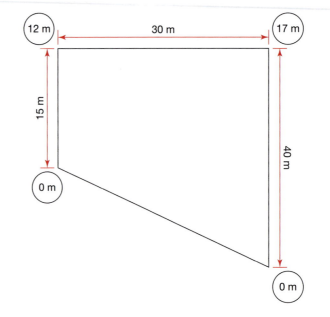

Figure 18.16

Answers to Exercise 18.1

1. 26.25 cm² (mid-ordinate rule); 26.25 cm² (trapezoidal rule)
2. 4485 m² (mid-ordinate rule); 4485 m² (trapezoidal rule); 4490 m² (Simpson's rule)
3. 93.3 cm²
4. (a) 154.8 m²; (b) 185.76 m³
5. 3324 m²
6. 10 656.48 m²
7. 5306.67 m³
8. 121 m³
9. 7829.25 m³
10. 6137.5 m³

Trigonometry (2)

Learning outcomes:

(a) Show, that in ΔABC:

$$\frac{a}{\sin A} = \frac{b}{\sin B} = \frac{c}{\sin C} \text{ (sine rule)}$$

$a^2 = b^2 + c^2 - 2bc \cos A$ (cosine rule)

(b) Apply the sine rule and the cosine rule to solve triangles and practical problems in construction

(c) Find the area of a triangle if two sides and the included angle are given

19.1 The sine rule and the cosine rule

In Chapter 13 we applied the trigonometrical ratios to determine the unknown angles and sides of right-angled triangles. For triangles without a right angle, the trigonometrical ratios cannot be applied directly; instead sine and cosine rules may be used to determine the unknown angles and sides.

19.1.1 The sine rule

The sine rule states that in any triangle the ratio of the length of a side to the sine of the angle opposite that side is constant:

$$\frac{a}{\sin A} = \frac{b}{\sin B} = \frac{c}{\sin C}$$

where a, b and c are the sides of ΔABC, as shown in Figure 19.1. Side a, b and c are opposite ∠A, ∠B and ∠C, respectively.

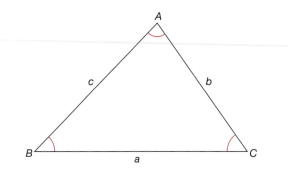

Figure 19.1

To prove the sine rule, draw a perpendicular from *A* to meet line *BC* at point *D* (Figure 19.2a). Triangles *ADB* and *ACD* are right-angled triangles; therefore we can apply the trigonometric ratios to determine the length of *AD*. In △*ADB*,

$$\frac{AD}{c} = \sin B$$

Therefore, $AD = c \sin B$ (1)

Similarly in △*ACD*, $AD = b \sin C$ (2)

From equations (1) and (2):

$$c \sin B = b \sin C$$

or $\dfrac{c}{\sin C} = \dfrac{b}{\sin B}$ (3)

In △*BAC*, draw a perpendicular from *B* to meet line *CA* at *E*, as shown in Figure 19.2b.

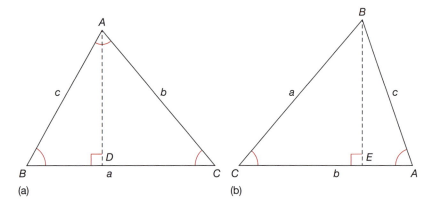

(a) (b)

Figure 19.2

In △*BAE*, $\dfrac{BE}{c} = \sin A$

Therefore, $BE = c \sin A$ (4)

Similarly in △*BEC*, $BE = a \sin C$ (5)

From equations (4) and (5):

$c \sin A = a \sin C$

or $\dfrac{c}{\sin C} = \dfrac{a}{\sin A}$ (6)

From equations (3) and (6):

$$\dfrac{a}{\sin A} = \dfrac{b}{\sin B} = \dfrac{c}{\sin C}$$

The above rule can be adapted if we have triangles *PQR*, *XYZ* and so on.
In $\triangle PQR$,

$$\dfrac{p}{\sin P} = \dfrac{q}{\sin Q} = \dfrac{r}{\sin R}$$

In $\triangle XYZ$,

$$\dfrac{x}{\sin X} = \dfrac{y}{\sin Y} = \dfrac{z}{\sin Z}$$

The sine rule may be used for the solution of triangles when:

1. two angles and one side are known

2. two sides and the angle opposite one of them are known.

Example 19.1

In $\triangle ABC$, $\angle B = 50°$, $\angle C = 75°$ and $AB = 80$ cm. Find $\angle A$ and sides BC and AC.

Solution:

$\triangle ABC$ is shown in Figure 19.3.

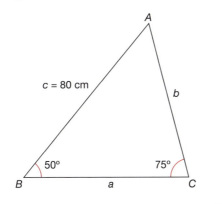

Figure 19.3

$\angle A = 180° - \angle B - \angle C$

$\quad = 180° - 50° - 75° = \mathbf{55°}$

To find side *c*, solve: $\dfrac{a}{\sin A} = \dfrac{c}{\sin C}$

Transposing, $a = \dfrac{c \times \sin A}{\sin C}$

$$= \dfrac{80 \times \sin 55°}{\sin 75°} \quad (c = \text{side } AB = 80 \text{ cm})$$

$$= \dfrac{80 \times 0.8192}{0.966} = \textbf{67.84 cm} = \textbf{side } \textbf{\textit{BC}}$$

$$\dfrac{b}{\sin B} = \dfrac{c}{\sin C}$$

Transposing, $b = \dfrac{c \times \sin B}{\sin C}$

$$= \dfrac{80 \times \sin 50°}{\sin 75°}$$

$$= \dfrac{80 \times 0.766}{0.966} = \textbf{63.44 cm} = \textbf{side } \textbf{\textit{AC}}$$

Example 19.2

The members of a roof truss slope at 35° and 60°, as shown in Figure 19.4. Calculate the length of all members of the truss.

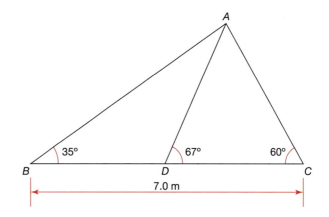

Figure 19.4

Solution:

In $\triangle ABC$:

$\quad \angle B = 35°, \angle C = 60°$

$\quad$ Therefore, $\angle A = 180 - 35° - 60° = 85°$

In $\triangle ABC$,

$$\dfrac{a}{\sin A} = \dfrac{b}{\sin B} = \dfrac{c}{\sin C}$$

$$\dfrac{7}{\sin 85°} = \dfrac{b}{\sin 35°} = \dfrac{c}{\sin 60°}$$

To calculate b, solve:

$$\frac{7}{\sin 85°} = \frac{b}{\sin 35°}$$

$$b = \frac{7 \times \sin 35°}{\sin 85°} = \textbf{4.03 m (AC)}$$

$$\frac{7}{\sin 85°} = \frac{c}{\sin 60°}$$

$$c = \frac{7 \times \sin 60°}{\sin 85°} = \textbf{6.085 m (AB)}$$

In $\triangle ADC$ (Figure 19.5a),

$$\frac{a}{\sin A} = \frac{c}{\sin C} = \frac{d}{\sin D}$$

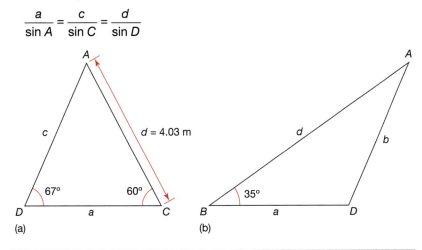

(a)   (b)

Figure 19.5

$$\angle A = 180° - 67° - 60° = \textbf{53°}$$

$$\frac{a}{\sin 53°} = \frac{c}{\sin 60°} = \frac{4.03}{\sin 67°}$$

$$a = \frac{4.03 \times \sin 53°}{\sin 67°} = \textbf{3.496 m (DC)}$$

$$c = \frac{4.03 \times \sin 60°}{\sin 67°} = \textbf{3.791 m (AD)}$$

In $\triangle ABD$ (Figure 19.5b), $\angle B = 35°$, $\angle D = 180° - 67 = 113°$, $\angle A = 32°$

$$\frac{a}{\sin A} = \frac{b}{\sin B}$$

$$\frac{a}{\sin 32°} = \frac{3.791}{\sin 35°}$$

$$a = \frac{3.791 \times \sin 32°}{\sin 35°} = \textbf{3.502 m (BD)}$$

19.1.2 The cosine rule

The cosine rule may be used for the solution of triangles, if:

1. two sides and the included angle are known
2. three sides are known

The cosine law can be written in the form of the following equations

$$a^2 = b^2 + c^2 - 2bc \cos A$$
$$b^2 = c^2 + a^2 - 2ac \cos B$$
$$c^2 = a^2 + b^2 - 2ab \cos C$$

To prove the cosine rule, consider $\triangle ABC$, as shown in Figure 19.6.

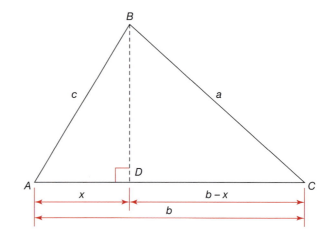

Figure 19.6

From B draw a perpendicular to meet line AC at D. In $\triangle ABD$,

$$(BD)^2 = c^2 - x^2 \tag{1}$$

Similarly in $\triangle BCD$

$$(BD)^2 = a^2 - (b - x)^2$$
$$= a^2 - b^2 + 2bx - x^2 \tag{2}$$

From equations (1) and (2):

$$c^2 - x^2 = a^2 - b^2 + 2bx - x^2$$
$$c^2 = a^2 - b^2 + 2bx \tag{3}$$

In $\triangle ABD$,

$$x = c \cos A \tag{4}$$

From equations (3) and (4):

$$c^2 \quad = a^2 - b^2 + 2bc \cos A$$
$$\text{or } a^2 = b^2 + c^2 - 2bc \cos A$$

Similarly, it can be shown that,

$$b^2 = c^2 + a^2 - 2ca \cos B$$
$$c^2 = a^2 + b^2 - 2ab \cos C$$

Example 19.3

Solve the triangle shown in Figure 19.7

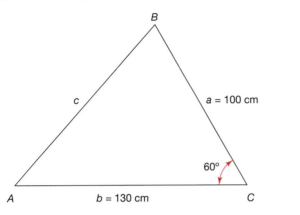

Figure 19.7

Solution:

In this question we have been asked to determine side c, $\angle A$ and $\angle B$.
 We have been given the lengths of the two sides and the included angle, therefore we'll use the cosine rule. As $\angle C$ is known, we must use the version of the cosine rule that contains cos C, i.e.,

$$c^2 = a^2 + b^2 - 2ab \cos C$$
$$= 100^2 + 130^2 - 2 \times 100 \times 130 \times \cos 60°$$
$$= 10\ 000 + 16\ 900 - 26\ 000 \times 0.5$$
$$= 26\ 900 - 13\ 000 = 13\ 900$$

Therefore, $c = \sqrt{13\ 900} = \textbf{117.898 cm}$

To find $\angle A$, use:

$$a^2 = b^2 + c^2 - 2bc \cos A$$
$$100^2 = 130^2 + 117.898^2 - 2 \times 130 \times 117.898 \times \cos A$$
$$10\ 000 = 16\ 900 + 13\ 900 - 30\ 653.48 \cos A$$
$$30\ 653.48 \cos A = 30\ 800 - 10\ 000$$
$$\cos A = 20\ 800/30\ 653.48 = 0.67855$$
$$\angle A = \cos^{-1} 0.67855 = \textbf{47°16'10''}$$
$$\angle B = 180° - 60° - 47°16'10'' = \textbf{72°43'50''}$$

Example 19.4

A large piece of land has been divided into triangles to facilitate the survey of the area. The dimensions of one of the triangles are shown in Figure 19.8.

(a) Identify the technique that can be used to determine the angles of the triangle.

(b) Calculate the value of the angles.

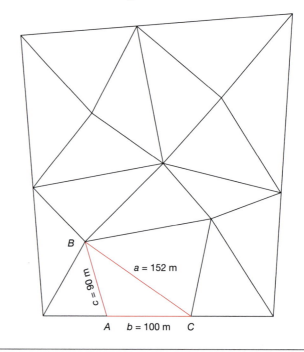

Figure 19.8

Solution:

(a) Initially the cosine rule may be used to find one of the angles. The second angle may be determined by either the cosine rule or the sine rule. The third angle may be calculated by subtracting the sum of the two angles already determined from 180°.

(b) According to the cosine rule,

$$a^2 = b^2 + c^2 - 2bc \cos A$$

$$152^2 = 100^2 + 90^2 - 2 \times 100 \times 90 \times \cos A$$

$$23\,104 = 10\,000 + 8100 - 18\,000 \cos A$$

$$18\,000 \cos A = -5004$$

$$\cos A = -5004 \div 18\,000 = -0.278$$

$$\angle A = \cos^{-1}(-0.278) = 106.14° \text{ or } \mathbf{106°08'27''}$$

To calculate $\angle C$, use

$$c^2 = a^2 + b^2 - 2ab \cos C$$

$$90^2 = 152^2 + 100^2 - 2 \times 152 \times 100 \times \cos C$$

8100 = 23 104 + 10 000 − 30 400 cos C

cos C = 25 004 ÷ 30 400 = 0.8225

$\angle C$ = cos⁻¹ 0.8225 = 34.664° or **34°39′51″**

$\angle B$ = 180° − 106°08′27″ − 34°39′51″ = **39°11′42″**

19.2 Area of triangles

If two sides and the included angle are known, then trigonometry may be used to determine the area of the triangle:

$$\text{Area} = \frac{1}{2}ab \sin C = \frac{1}{2}bc \sin A = \frac{1}{2}ca \sin B$$

Depending on the information given, one of the above formulae may be used for area calculation.

Example 19.5

Find the area of the triangle shown in Figure 19.7

Solution:

In Figure 19.7, a = 100 cm; b = 130 cm; angle between sides a and b, i.e. $\angle C$ = 60°

$$\text{Area of } \triangle ABC = \frac{1}{2}ab \sin C$$

$$= \frac{1}{2}100 \times 130 \times \sin 60°$$

$$= \frac{1}{2}100 \times 130 \times 0.866$$

$$= \textbf{5629 cm}^2$$

Example 19.6

The plan of a building plot is shown in Figure 19.9. Calculate:

(a) $\angle D$, $\angle B$, $\angle DAC$, $\angle DCA$ and $\angle CAB$

(b) the area of the building plot.

Solution:

(a) Consider $\triangle ACB$ and apply the sine rule:

$$\frac{a}{\sin A} = \frac{b}{\sin B} = \frac{c}{\sin C}$$

$$\frac{50}{\sin A} = \frac{b}{\sin B} = \frac{55}{\sin 50}$$

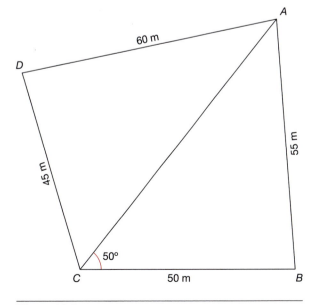

Figure 19.9

To find $\angle A$, solve:

$$\frac{50}{\sin A} = \frac{55}{\sin 50}$$

$$50 \times \sin 50 = 55 \times \sin A$$

$$\sin A = \frac{50 \times \sin 50}{55} = 0.6964$$

$\angle A$ (or $\angle CAB$) = $\sin^{-1} 0.6964$ = **44°08′21″**

$\angle B = 180° - 50° - 44°08′21″$ = **85°51′39″**

To find the length of AC, use the sine rule:

$$\frac{b}{\sin B} = \frac{c}{\sin C}$$

$$\frac{b}{\sin 85°\,51′\,39″} = \frac{55}{\sin 50°}$$

$$b = \frac{55 \times \sin 85°\,51′\,39″}{\sin 50°} = \textbf{71.6 m}$$

To find the angles of $\triangle ADC$, use the cosine rule. According to the cosine rule:

$$a^2 = c^2 + d^2 - 2cd \cos A$$

$$45^2 = 60^2 + 71.6^2 - 2 \times 60 \times 71.6 \times \cos A$$

$$2025 = 3600 + 5126.56 - 8592 \cos A$$

$$8592 \cos A = 6701.56$$

$$\cos A = 6701.56 \div 8592 = 0.7799$$

$\angle A$ (or $\angle DAC$) = $\cos^{-1} 0.7799$ = 38.74° or **38°44′30″**

To find $\angle D$, use the sine rule, i.e.

$$\frac{a}{\sin A} = \frac{d}{\sin D}$$

$$\sin D = \frac{71.6 \times \sin 38°\,44′\,30″}{45} = 0.9957333$$

$\angle D$ = **84°42′19″**

$\angle DCA = 180° - 38°44′30″ - 84°42′19″$ = **56°33′11″**

(b)

Area of the plot = area of $\triangle ACB$ + area of $\triangle ADC$

$$= \frac{1}{2}\,50 \times 71.6 \times \sin 50° + \frac{1}{2}\,45 \times 71.6 \times \sin 56°33′11″$$

$$= 1371.22 + 1344.21$$

$$= \textbf{2715.43 m}^2$$

Exercise 19.1

The solutions to Exercise 19.1 can be found in Appendix 2.

1. In $\triangle ABC$, $\angle A = 72°$, sides a and c are both equal to 50 cm. Calculate side b, $\angle B$ and $\angle C$.

2. In $\triangle ABC$, $\angle B = 40°$, $\angle C = 55°$ and $a = 30$ cm. Find $\angle A$ and sides b and c.

3. A Fink truss is shown in Figure 19.10. Find the length of members AE, EC, AG, GE and FG.

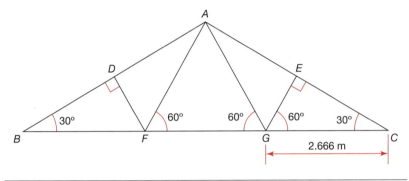

Figure 19.10

4. In $\triangle PQR$, $\angle Q = 44°$, side $p = 30$ cm and side $r = 25$ cm. Calculate $\angle P$, $\angle R$ and side q.

5. In $\triangle ABC$, side $a = 8$ cm, side $b = 6$ cm and side $c = 10$ cm. Calculate $\angle A$, $\angle B$ and $\angle C$.

6. A surveyor set up his instrument at point A and took measurements of a building plot as shown in Figure 19.11. Calculate the area of the plot.

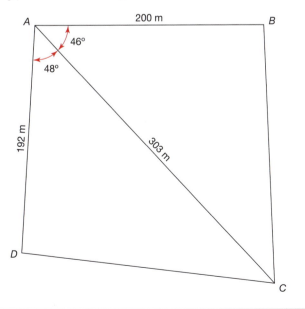

Figure 19.11

7. The plan of a building plot is shown in Figure 19.12. Calculate $\angle A$, $\angle C$, $\angle ABC$ and $\angle ADC$.

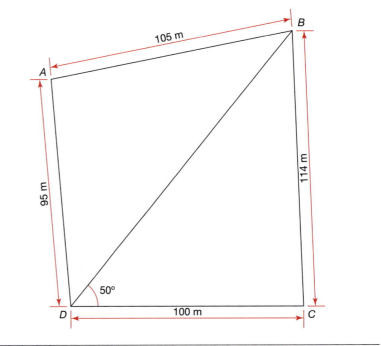

Figure 19.12

8. A surveyor takes measurements of a building plot from two corners, as shown in Figure 19.13. Calculate the length of sides AD, AB and BC.

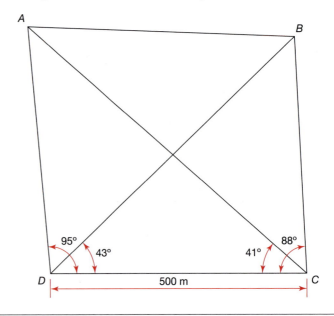

Figure 19.13

9. Figure 19.14 shows the plan of a plot of land. Find ∠*HEF*, ∠*EFG*, sides *EH*, *EF* and *FG*.

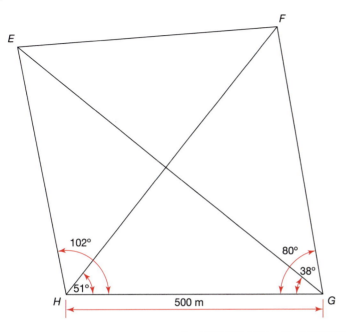

Figure 19.14

Answers to Exercise 19.1

1. Side *b* = 30.9 cm; ∠*B* = 36°; ∠*C* = 72°

2. ∠*A* = 85°; *b* = 19.36 cm; *c* = 24.67 cm

3. *AE* = 2.309 m; *EC* = 2.309 m; *AG* = 2.666 m; *GE* = 1.333 m;
   FG = 2.666 m

4. ∠*P* = 80.681°; ∠*R* = 55.319°; side *q* = 21.12 cm

5. ∠*A* = 53.13°; ∠*B* = 36.87°; ∠*C* = 90°

6. 43 412.59 m²

7. ∠*A* = 95.936°; ∠*C* = 87.781°
   ∠*ABC* = 81.671°; ∠*ADC* = 94.6126°

8. *AD* = 472.217 m; *AB* = 525.726 m; *BC* = 451.828 m

9. ∠*HEF* = 82.328° (82°19′40.8″); ∠*EFG* = 95.672° (95°40′19.2″);
   EH = 478.899 m; *EF* = 511.623 m; *FG* = 514.864 m

CHAPTER 20

Computer techniques

Learning outcomes:

(a) Perform simple calculations involving addition, subtraction, multiplication and division

(b) Use a range of functions to determine the sum, mean, range, etc.

(c) Display the results as pie charts and bar charts

20.1 Introduction

We have at our disposal many devices like calculators and computers to make mathematical calculations easier and faster than the manual techniques. It has taken inventors several centuries to reach the present stage. In the early 1600s John Napier, a Scottish mathematician, invented logarithms. The use of logarithms made several types of calculations easier and faster. Later, William Oughtred used Napier's logarithms as the basis for the slide rule, which remained in common use for over 300 years. The use of the slide rule came to an end in the early 1970s, when pocket calculators became popular. The next development was an electronic spreadsheet invented by Daniel Bricklin, a student at Harvard Business School. Originally the electronic spreadsheet was invented to make calculations in business easier, but now it is used in many disciplines. Many spreadsheet softwares are available now, but this chapter is based on the widely used Microsoft Excel.

20.2 Microsoft Excel 2000

Microsoft Excel 2000 is a spreadsheet program that lets us work with numbers and text. An Excel file is known as a workbook, and one workbook can hold several sheets, e.g. sheet 1, sheet 2, sheet 3, etc. Each sheet is divided into rows and columns and their intersections create cells, which are known by a reference. A cell reference is a combination of letter(s) and number, e.g. B10 or AA5 to signify the intersection of a column and a row. Figure 20.1 shows a spreadsheet in which the reference of the selected cell is D6.

Chart wizard

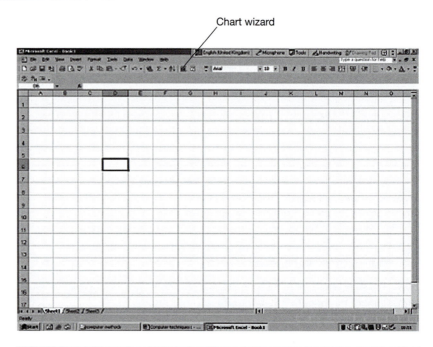

Figure 20.1

Each cell can hold text, a number or a formula. A formula is a special way to tell Excel to perform a calculation using information present in other cells. Formulae can use not only the normal arithmetic operations like plus (+), minus (–), multiply (*) and divide (/), but also the built-in functions to find square roots, average, sine of an angle, etc. A summary of these is given in Table 20.1.

Table 20.1

Symbol/function	Action
+	Addition
–	Subtraction
*	Multiplication
/	Division
=	Equal to
^	Raised to a power
Sum	The sum of the values
Average	The average (or mean) of the values

Example 20.1

Jas is planning to renovate her house and wants to replace the old carpets with laminated flooring in two rooms. If the rooms measure

5.8 m × 4.2 m and 4.5 m × 4.1 m, and one pack of the laminated boards covers 2.106 m^2, find the number of packs required. Consider wastage as 10%.

Solution:

The cells are labelled as length, width, area, etc. on a Microsoft Excel spreadsheet, as shown in Figure 20.2. The length and the width of Room 1 are entered and multiplied to determine the floor area. The formula for the floor area (= B4*C4) is entered in cell D4. It is necessary to use the equal sign before a formula, otherwise the computer will not do the required calculation. The area (cell D4) is divided by 2.106 to calculate the number of packs required, as shown in cell E4. Cell F4 shows the formula to work out the wastage, 10/100 representing 10%. The total number of packs required (cell G4) is the sum of the answers in cells E4 and F4, i.e. the number of actual packs required plus the wastage.

Enter the dimensions of Room 2, and replicate all the formulae. The total number of packs is the sum of cells G4 and G6. Figure 20.2a shows the formulae used and Figure 20.2b the results.

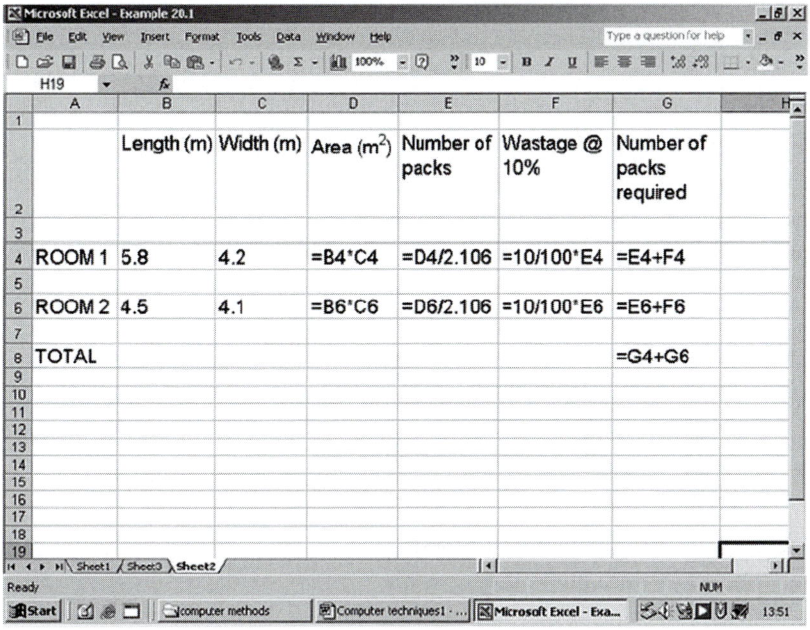

	A	B	C	D	E	F	G
1							
2		Length (m)	Width (m)	Area (m^2)	Number of packs	Wastage @ 10%	Number of packs required
3							
4	ROOM 1	5.8	4.2	=B4*C4	=D4/2.106	=10/100*E4	=E4+F4
5							
6	ROOM 2	4.5	4.1	=B6*C6	=D6/2.106	=10/100*E6	=E6+F6
7							
8	TOTAL						=G4+G6

Figure 20.2a

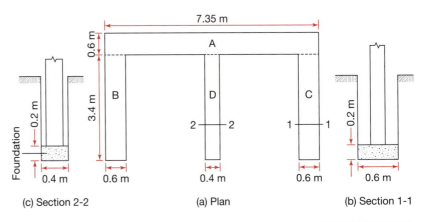

Figure 20.2b

The spreadsheet shows:

	A	B	C	D	E	F	G
1		Length (m)	Width (m)	Area (m²)	Number of packs	Wastage @ 10%	Number of packs required
4	ROOM 1	5.8	4.2	24.36	11.57	1.157	12.72
6	ROOM 2	4.5	4.1	18.45	8.76	0.876	9.64
8	TOTAL						22.36
10					The answer is 23 packs		

Example 20.2

Figure 20.3 shows the foundation plan of a building extension and the sectional details of the foundation. Find the volume of concrete required to build a 0.200 m thick foundation.

(c) Section 2-2 (a) Plan (b) Section 1-1

Figure 20.3

Solution:

The foundation plan is divided into four parts: A, B, C and D. The length and width are entered in columns B and C. The thickness of the foundation (0.200 m) is uniform throughout and entered in cells D4 to D7. The volume of concrete is determined by multiplying the length, width and height (or thickness); therefore, the formula entered in cell E4 is: =B4*C4*D4. The formula is replicated in cells E5, E6 and E7, as shown in Figure 20.4a. Figure 20.4b shows the results.

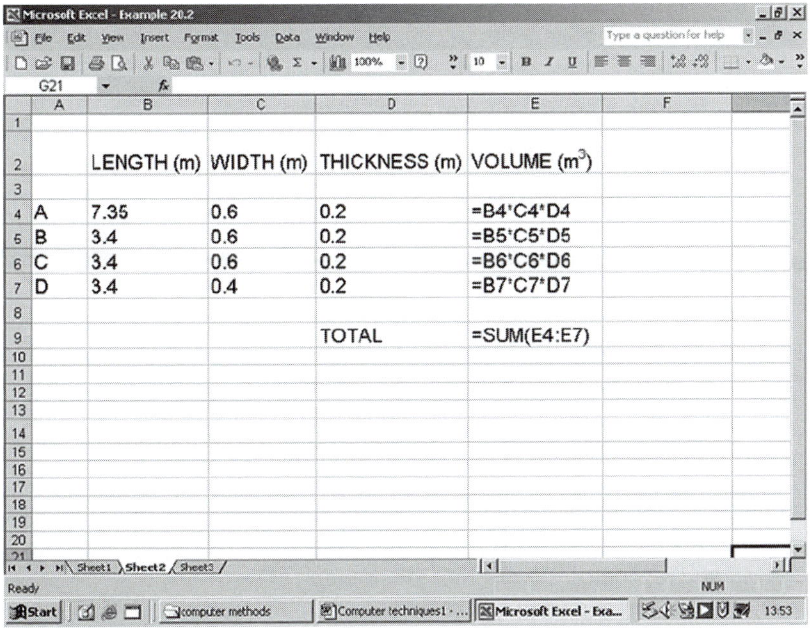

	A	B	C	D	E	F
1						
2		LENGTH (m)	WIDTH (m)	THICKNESS (m)	VOLUME (m³)	
3						
4	A	7.35	0.6	0.2	=B4*C4*D4	
5	B	3.4	0.6	0.2	=B5*C5*D5	
6	C	3.4	0.6	0.2	=B6*C6*D6	
7	D	3.4	0.4	0.2	=B7*C7*D7	
8						
9				TOTAL	=SUM(E4:E7)	

Figure 20.4a

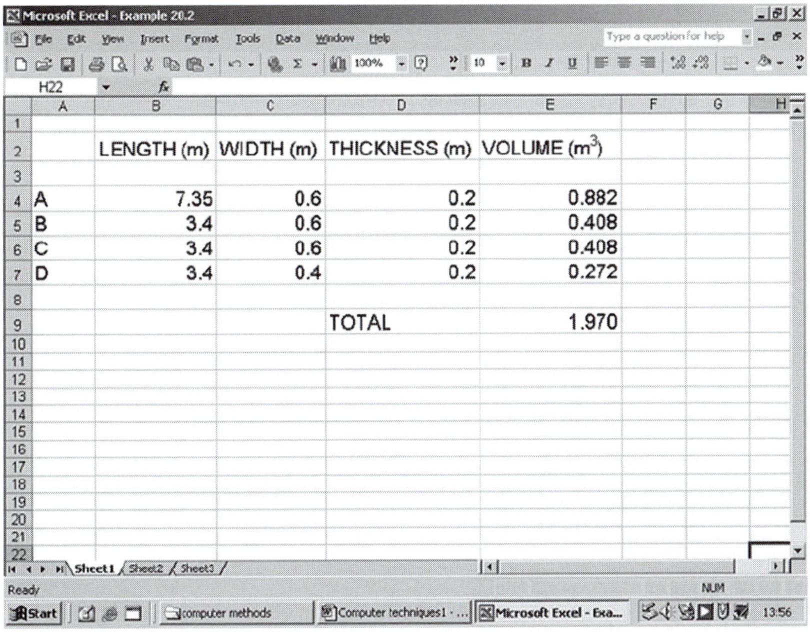

	A	B	C	D	E	F	G
1							
2		LENGTH (m)	WIDTH (m)	THICKNESS (m)	VOLUME (m³)		
3							
4	A	7.35	0.6	0.2	0.882		
5	B	3.4	0.6	0.2	0.408		
6	C	3.4	0.6	0.2	0.408		
7	D	3.4	0.4	0.2	0.272		
8							
9				TOTAL	1.970		

Figure 20.4b

Example 20.3

Five groups of students were asked to prepare concrete cubes and test them to failure to determine the crushing strength. The test results were:

Group 1: 50, 52, 53, 53, 55, 57, 60, 54, 54, 55 N/mm^2

Group 2: 50, 52, 53, 54, 55, 58, 61, 54, 54, 55 N/mm^2

Group 3: 49, 52, 52, 53, 55, 58, 60, 55, 54, 56 N/mm^2

Group 4: 50, 52, 53, 60, 55, 59, 62, 54, 54, 55 N/mm^2

Group 5: 48, 52, 52, 53, 55, 57, 58, 54, 53, 55 N/mm^2

Find the mean and the range of each group's data.

Solution:

The data of all the groups are entered as shown in Figure 20.5a. To find the mean, the formula is: =average(first cell:last cell). For Group 1 this is: =average(B3:K3). The formula is entered without any spaces. The range of a set of data is equal to the maximum value minus the minimum value. Max(B3:K3) means the maximum value from the cells ranging between B3 and K3. Similarly, Min(B3:K3) means the minimum value from the cells ranging between B3 and K3. The formula to find the range is entered in cell M3 and replicated in cells M4, M5, M6 and M7.

The results are shown in Figure 20.5b.

A	B	C	D	E	F	G	H	I	J	K	L	M
Group	Crushing strength (N/mm^2)										Mean	Range
Group 1	50	52	53	53	55	57	60	54	54	55	=AVERAGE(B3:K3)	=MAX(B3:K3)-MIN(B3:K3)
Group 2	50	52	53	54	55	58	61	54	54	55	=AVERAGE(B4:K4)	=MAX(B4:K4)-MIN(B4:K4)
Group 3	49	52	52	53	55	58	60	55	54	56	=AVERAGE(B5:K5)	=MAX(B5:K5)-MIN(B5:K5)
Group 4	50	52	53	60	55	59	62	54	54	55	=AVERAGE(B6:K6)	=MAX(B6:K6)-MIN(B6:K6)
Group 5	48	52	52	53	55	57	58	54	53	55	=AVERAGE(B7:K7)	=MAX(B7:K7)-MIN(B7:K7)

Figure 20.5a

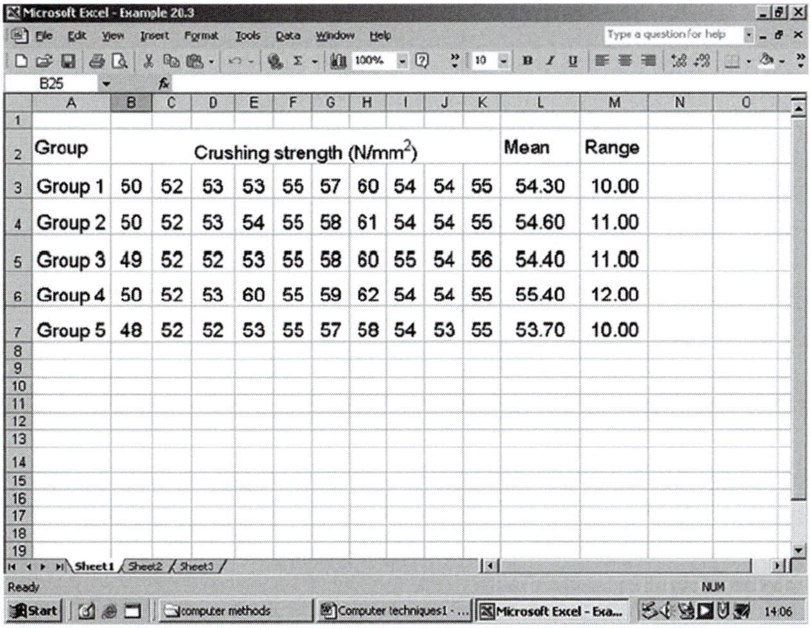

Figure 20.5b

Example 20.4

Twenty-six bricks were tested to determine their compressive strength in N/mm^2. The results are:

| 49 | 50 | 55 | 54 | 51 | 52 | 56 | 55 | 53 | 54 | 54 | 53 | 60 |
| 55 | 53 | 58 | 61 | 56 | 57 | 52 | 54 | 57 | 55 | 58 | 56 | 59 |

Arrange the data into five groups and calculate the mean compressive strength.

Solution:

The data have been arranged into five groups as shown in Figure 20.6a. The class midpoints are entered in cells C4 to C8. As the mean of grouped data is determined by the formula: Mean $= \dfrac{\Sigma fx}{\Sigma f}$, the next step is to find the product of frequency (f) and class midpoint (x). The formula to determine fx is entered in cell D4 and replicated for the other cells. The sum of the frequencies (Σf) and the sum of fx (Σfx) is found by entering the formulae shown in Figure 20.6a. Finally, the mean is shown in cell E11.

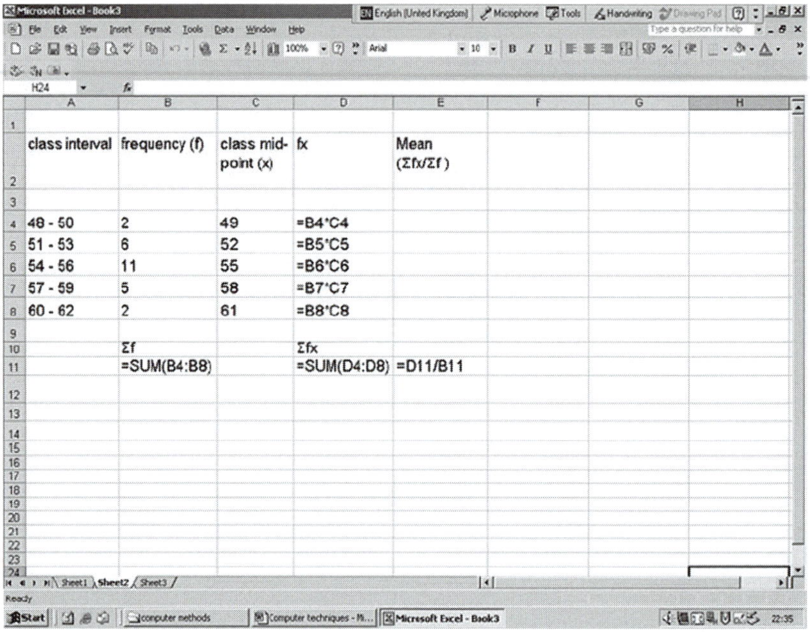

	A	B	C	D	E
1					
2	class interval	frequency (f)	class mid-point (x)	fx	Mean (Σfx/Σf)
3					
4	48 - 50	2	49	=B4*C4	
5	51 - 53	6	52	=B5*C5	
6	54 - 56	11	55	=B6*C6	
7	57 - 59	5	58	=B7*C7	
8	60 - 62	2	61	=B8*C8	
9					
10		Σf		Σfx	
11		=SUM(B4:B8)		=SUM(D4:D8)	=D11/B11

Figure 20.6a

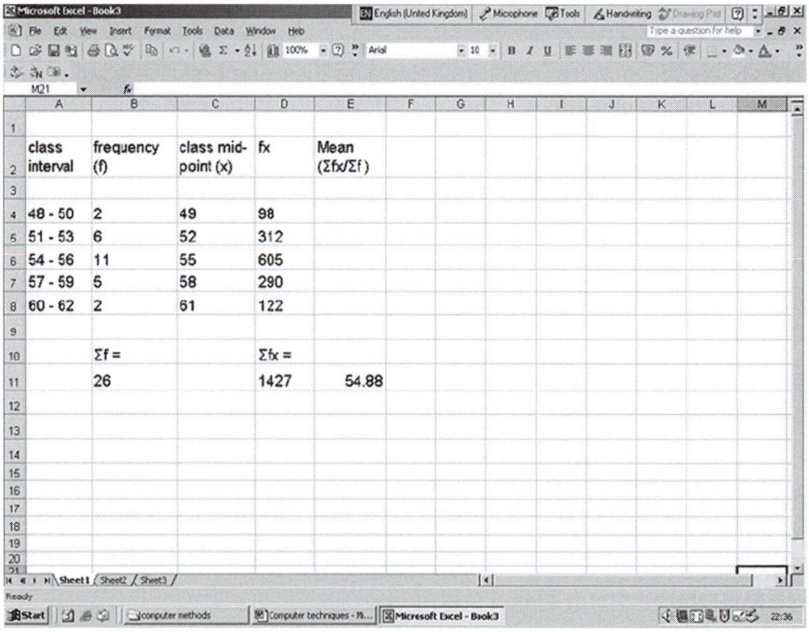

	A	B	C	D	E
1					
2	class interval	frequency (f)	class mid-point (x)	fx	Mean (Σfx/Σf)
3					
4	48 - 50	2	49	98	
5	51 - 53	6	52	312	
6	54 - 56	11	55	605	
7	57 - 59	5	58	290	
8	60 - 62	2	61	122	
9					
10		Σf =		Σfx =	
11		26		1427	54.88

Figure 20.6b

Example 20.5

The number of personnel in S & R Consulting Engineers is given in the table:

Directors	Associate directors	Civil engineers	Technician engineers	Technicians	Administrative staff
2	4	3	6	4	3

Represent this information in the form of: (a) a horizontal bar chart; (b) a pie chart.

Solution:

The data are entered into the cells, as shown in Figure 20.7a. The data and the headings are selected and the chart wizard icon clicked to display the list of charts available. The horizontal bar chart is selected and the computer instructions followed to produce the diagram.

To produce the pie chart (Figure 20.7b) an appropriate type of pie is selected and instructions that appear on the computer screen followed.

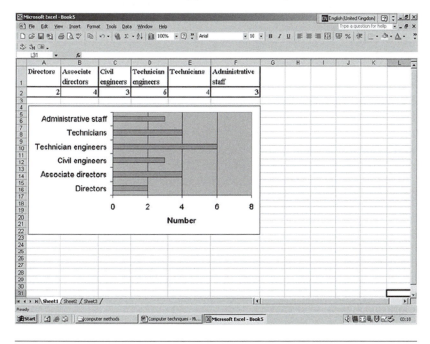

Figure 20.7a

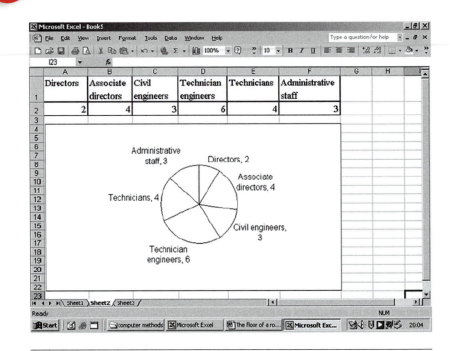

Figure 20.7b

Example 20.6

The heat loss from a room is given by:

Heat loss = $U \times A \times T$

where U is the U-value of the material/component, A is the surface area in m², T is the temperature difference in °C (or K).

If the temperature difference between the inside and outside air is 20°C, find the heat loss from a room given that:

U-values (W/m² °C)

Cavity wall = 0.35; floor = 0.35; roof = 0.30; door = 0.46; window = 2.6

Areas (m²)

Wall = 46.5 (gross); floor = 15; roof = 15; door = 1.7; window = 2.0

Solution:

The first step is to find the net wall area, which can be determined by taking away the area of door and window from the gross wall area. The U-values, areas and the temperature difference are entered into a worksheet and multiplied, as the heat loss = $U \times A \times T$. The heat losses from the various components are added to find the total heat loss, as shown in Figure 20.8.

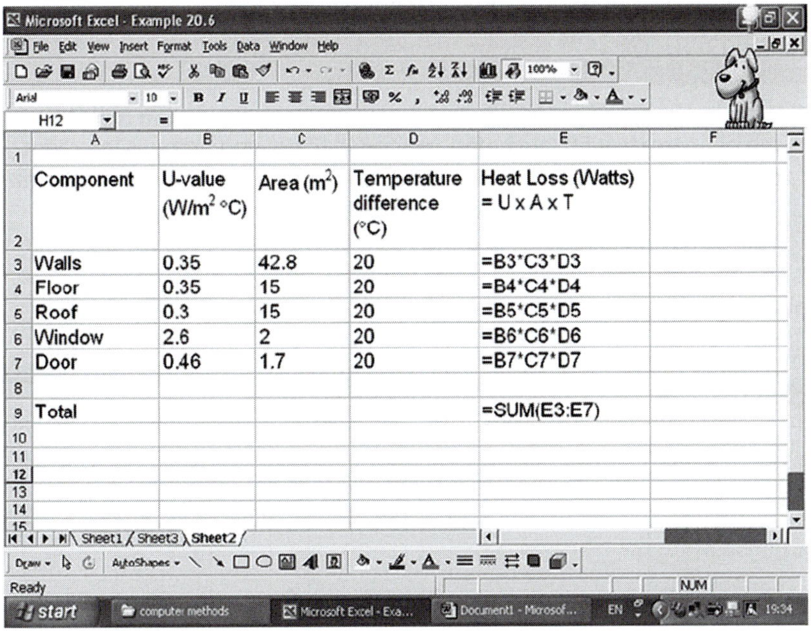

	A	B	C	D	E	F
1						
2	Component	U-value (W/m^2 °C)	Area (m^2)	Temperature difference (°C)	Heat Loss (Watts) = U x A x T	
3	Walls	0.35	42.8	20	=B3*C3*D3	
4	Floor	0.35	15	20	=B4*C4*D4	
5	Roof	0.3	15	20	=B5*C5*D5	
6	Window	2.6	2	20	=B6*C6*D6	
7	Door	0.46	1.7	20	=B7*C7*D7	
8						
9	Total				=SUM(E3:E7)	

Figure 20.8a

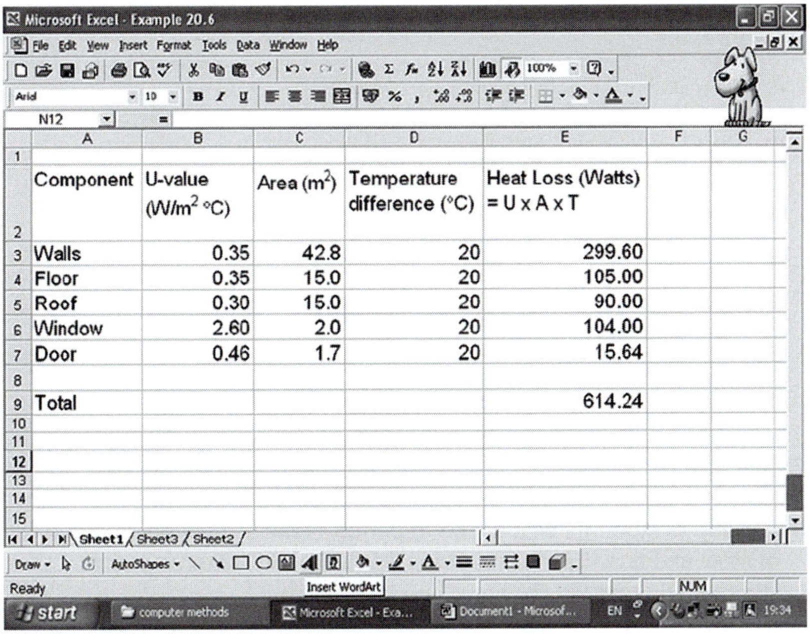

	A	B	C	D	E	F	G
1							
2	Component	U-value (W/m^2 °C)	Area (m^2)	Temperature difference (°C)	Heat Loss (Watts) = U x A x T		
3	Walls	0.35	42.8	20	299.60		
4	Floor	0.35	15.0	20	105.00		
5	Roof	0.30	15.0	20	90.00		
6	Window	2.60	2.0	20	104.00		
7	Door	0.46	1.7	20	15.64		
8							
9	Total				614.24		

Figure 20.8b

Exercise 20.1

1. Jane is planning to renovate her house and has decided to replace the old carpets with laminated flooring in the dining room, living room and the hall. The dining room measures 3.5 m × 3.5 m, the living room 6.1 m × 4.2 m and the hall 4 m × 2 m. If one pack of the laminated boards covers 2.106 m², find the number of packs required. Consider wastage as 10%.

2. Figure 20.9 shows the foundation plan of a building extension and the sectional details of the foundation. Find the volume of concrete required to build the 0.750 m deep foundation.

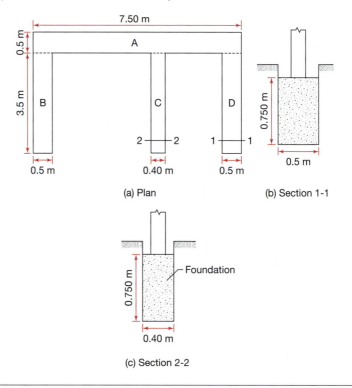

(a) Plan　　　　(b) Section 1-1

(c) Section 2-2

Figure 20.9

3. A group of students was asked to perform tensile tests on 15 samples of steel and find the maximum tensile force in kN taken by each sample. The test results were:

   9.9, 12.1, 10.0, 12.0, 10.2, 11.9, 10.3, 11.6, 10.5, 11.5, 11.3, 10.9, 11.3, 11.6, 11.0 kN.

   Find the mean tensile force and the range of the data.

4. Twenty-five samples of PVC were tested to determine their coefficient of linear thermal expansion. The results were:

63	70	66	70	74	71	67	77	78	72
66	76	64	69	73	82	69	76	70	65
71	72	74	78	73 $(\times 10^{-6}/°C)$					

Arrange the data into a number of appropriate groups and calculate the mean coefficient of thermal expansion.

5. The number of personnel in R & S Consulting Engineers is given in the table:

Directors	Associate directors	Civil engineers	Technician engineers	Technicians	Administrative staff
3	5	4	8	6	4

Represent this information in the form of a horizontal and a vertical bar chart.

6. The amounts of the materials required to make a concrete mix are: cement: 60 kg; sand: 100 kg; gravel: 200 kg; water: 35 kg. Represent the above information as a pie chart.

7. The heat loss from a room is given by:

Heat loss = $U \times A \times T$

If the temperature difference between the inside and outside air is 20°C, find the heat loss from a room given that:

U-values (W/m²°C)

Cavity wall = 0.35; floor = 0.35; roof = 0.30; door = 0.46;
        patio door = 2.6

Areas (m²)

Wall = 54.0 (gross); floor = 20; roof = 20; door = 1.7; patio door = 5.0

Answers to Exercise 20.1

1. 24
2. 6.488 m³
3. Mean = 11.07 N; range = 2.2 N
4. 71.4 × 10⁻⁶/°C
5. See Appendix 2
6. See Appendix 2
7. 866.74 W

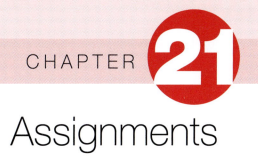

CHAPTER **21**

Assignments

21.1 Assignment 1

Task 1

(a) Find the area of the shapes shown in Figure 21.1.

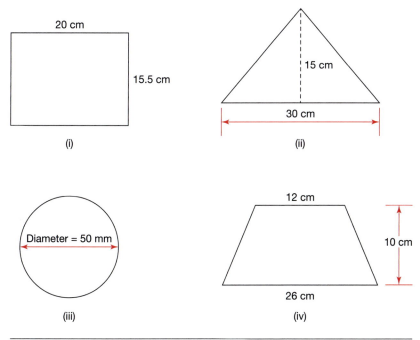

Figure 21.1

(b) Find the volume of the objects shown in Figure 21.2.

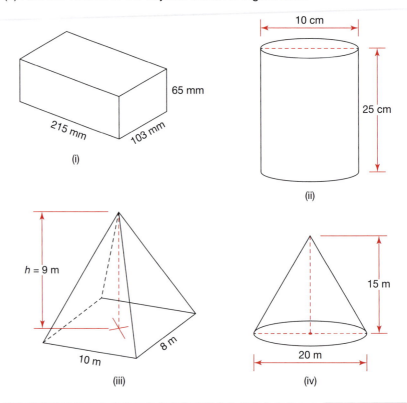

Figure 21.2

Task 2

Figure 21.3 shows the foundation and floor plan of a building. Calculate:

(a) the volume of soil that will be excavated to form 1000 mm deep foundation trenches

(b) the volume of the excavated soil after bulking, if the bulking factor is 10%

(c) the volume of 1:3:6 concrete required to construct 200 mm thick strip foundation

(d) the number of floor tiles required if each tile measures 330 mm × 330 mm. Allow 10% for wastage.

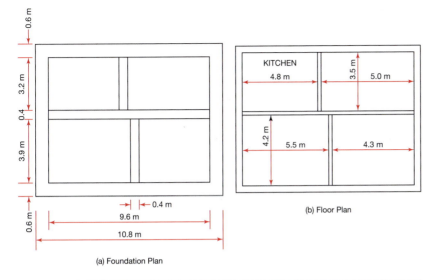

(a) Foundation Plan

(b) Floor Plan

Figure 21.3

Task 3

Figure 21.4 shows the cross-section of a drain and the depth of water flow. If the diameter of the drain is 300 mm, find:

(a) the cross-sectional area of the water flow

(b) the flow rate (m³/s) if the velocity of flow is 1.5 m/s.

Note: Velocity (m/s) = $\dfrac{\text{Flow rate (m}^3\text{/s)}}{\text{Cross-sectional area of flow (m}^2\text{)}}$

Figure 21.4

Task 4

Each step of a stair has a rise of 210 mm and a going of 230 mm. Find the pitch of the stair

Task 5

A surveyor who wants to find the height of a building sets up his instrument at point *A* and finds that the angle of elevation to the top of the building is 45°. He then moves the instrument by 40.0 m towards the building and finds that the angle of elevation now is 60°. Assuming that the ground is horizontal and the height of the instrument is 1.45 m, calculate the height of the building.

Task 6

In an experiment to prove Hooke's Law the following data was obtained by stretching a metal specimen:

Stretching force (kN)	0	1.5	3.0	4.5	6.0	7.5
Extension (mm)	0	0.02	0.04	0.06	0.08	0.10

(a) Plot a graph of stretching force (*F*) versus extension (*E*) and determine the law connecting *F* and *E*.
(b) Find the extension when the stretching force is increased to 8.5 kN, assuming the straight-line law still holds good.

21.2 Assignment 2

Task 1

(a) Use your scientific calculator to answer the following and show your answers to two decimal places:

   (i) $\dfrac{12.3 \times 280.9}{115.8}$

   (ii) $6.5^2 + 9.3^3$

   (iii) $(5.7 \times 10^6) \times (2.3 \times 10^{-4})$

   (iv) $\sin 50.5°$

(b) Use your scientific calculator to answer the following and show your answer to three significant figures:

   (i) $\sqrt{27.1} \times \sqrt{15.08}$

   (ii) $\dfrac{\sin 80.7°}{\cos 78.3°}$

   (iii) $\log_{10} 1575$

   (iv) $0.0055 \times 10^{2/3}$

Task 2

Solve graphically:

(a) the simultaneous equations, $x - y = -3$ and $2x + 3y = 4$

(b) the quadratic equation, $x^2 - 3x - 4 = 0$.

Task 3

The width of a rectangle is 5.0 cm shorter than its length. If the perimeter of the rectangle is 74.0 cm, find its length and width.

Task 4

The dimensions of a kitchen are shown in Figure 21.5. Calculate:

(a) the length of skirting board required

(b) the number of porcelain floor tiles measuring 600 mm × 600 mm, assuming wastage of 10%

(c) the quantity (in litres) of emulsion paint required for painting the walls. One litre of emulsion covers 10 m² (approx.).

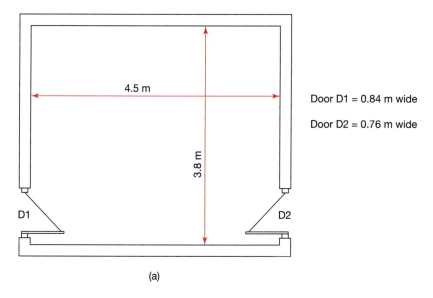

Door D1 = 0.84 m wide

Door D2 = 0.76 m wide

(a)

Figure 21.5 *(Continued)*

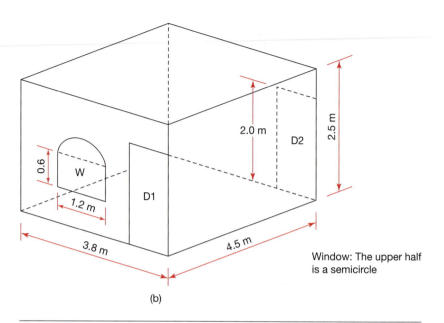

(b)

Figure 21.5 *(Continued)*

Task 5

Figure 21.6 shows the cross-section of a concrete drainage channel. Find the volume and mass of concrete used to make a 2.0 m long section. Assume the density of concrete to be 2400 kg/m³.

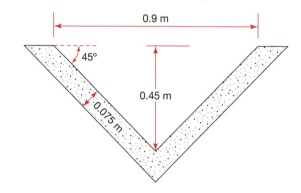

Figure 21.6

Task 6

A building is shaped like the frustum of a pyramid. If the building measures 75 m × 75 m at the base, 40 m × 40 m at the top and its vertical height is 40 m, calculate:

(a) the volume of space enclosed by the building

(b) the surface area, excluding the top and bottom.

Task 7

A hot water cylinder has a diameter of 500 mm and is 1500 mm high. The main body is a cylinder, but the top part is a hemisphere. Calculate:

(a) the capacity of the cylinder in litres (volume of a sphere $= \dfrac{4}{3}\pi r^3$; r is the radius)

(b) the surface area of the cylinder (including the base).

Task 8

Calculate the area of the plot of land shown in Figure 21.7 by two methods.

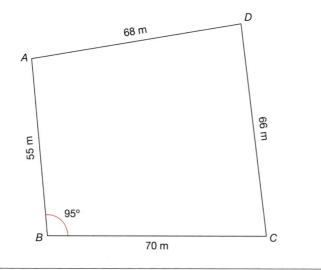

Figure 21.7

Task 9

The cross-sectional areas of a trench at 10 m intervals are shown in Figure 21.8. Use the trapezoidal rule and Simpson's rule to calculate the volume of earth excavated.

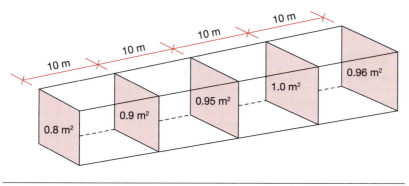

Figure 21.8

21.3 Assignment 3

Task 1

Bill Gateway has a 6.0 m long ladder that he wants to use to clean the rainwater gutters of his house. He knows that for safety the ladder should make an angle of 75° with the horizontal. How far from the wall should the foot of the ladder be?

Task 2

A surveyor wants to find the width of a river and stands on one bank at point C opposite a building (B), as shown in Figure 21.9. She sets up one station at point C and another at point A, which is 50 m from point C along the bank. On measuring the angles she finds $\angle BCA$ and $\angle CAB$ to be 85° and 60°, respectively. Find the width of the river.

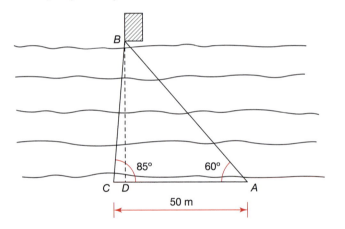

Figure 21.9

Task 3

Two technicians performed tests on bricks from the same batch to determine their compressive strength:

> Technician A: 30, 31, 35, 36, 38, 36, 39, 37, 34, 36 N/mm²
> Technician B: 32, 34, 36, 34, 37, 34, 35, 38, 37, 35 N/mm²

For each set of data, calculate:

(a) the mean, mode and median strength

(b) the range.

Compare the results and give your comments.

Task 4

The crushing strengths (unit: N/mm^2) of 40 concrete cubes are given here. Group the data into six or seven classes and find:

(a) the frequency of each class

(b) the mean crushing strength.

40	45	39	35	38	46	45	44	45	35
39	38	35	37	39	43	48	37	42	50
39	46	41	44	41	51	42	47	49	36
47	48	49	38	44	44	43	51	41	37

Task 5

Use the data given in Task 4 to produce a cumulative frequency curve, and find:

(a) the median crushing strength

(b) the interquartile range

(c) the number of samples having strength more than 40 N/mm^2.

1

Concrete mix

A1.1 1:2:4 concrete

A concrete mix can be prepared by two methods: (1) mixing by volume and (2) mixing by mass. In the latter method the mass of cement, fine aggregates (sand) and coarse aggregates (gravel, crushed stone, crushed bricks, blast-furnace slag, etc.) are calculated for preparing the concrete mix. In this section only the former method is described.

If we want to prepare 1 m^3 of 1:2:4 concrete mix, the quantities of dry materials may be calculated as shown below. The volume of cement in a 25 kg bag is 0.0166 m^3.

Assuming the volume of water to be 55% of the volume of cement, the proportions of the concrete mix may be written as 1:2:4:0.55. The total of these proportions is 7.55.

Cement : fine aggregates : coarse aggregates : water
1   : 2   : 4   : 0.55

$$\text{Volume of cement} = \frac{1}{7.55} \times 1 = 0.132 \text{ m}^3$$

$$\text{Volume of fine aggregates} = \frac{1}{7.55} \times 2 = 0.265 \text{ m}^3$$

$$\text{Volume of coarse aggregates} = \frac{1}{7.55} \times 4 = 0.53 \text{ m}^3$$

$$\text{Volume of water} = \frac{1}{7.55} \times 0.55 = 0.073 \text{ m}^3 \text{ or 73 litres}$$

The coarse aggregates contain a lot of air voids which are filled by cement, water and fine aggregates when they are mixed. The mass of each ingredient can be calculated by multiplying its volume by density:

Mass of 0.132 m^3 of cement = 0.132 × 1506 = 198.8 kg

Mass of 0.265 m^3 of sand = 0.265 × 1650 = 437.25 kg

Mass of 0.53 m^3 of gravel = 0.53 × 1600 = 848 kg

Mass of 0.073 m^3 of water = 0.073 × 1000 = 73 kg

Total = 1557 kg/m^3

The density of fresh concrete is 2350 kg/m^3.

The ratio in which the volumes of the dry materials should be increased is $\frac{2350}{1557}$ or 1.5. The volumes of dry materials that will give 1 m^3 of concrete after mixing are:

Volume of cement = 0.132 × 1.5 = 0.198 m^3

Volume of sand = 0.265 × 1.5 = 0.397 m^3

Volume of gravel = 0.53 × 1.5 = 0.8 m^3

Volume of water = 0.073 × 1.5 = 0.109 m^3 or 109 litres

A1.2 1:3:6 concrete

Cement	:	fine aggregates	:	coarse aggregates	:	water
1	:	3	:	6	:	0.55

Volume of cement = $\frac{1}{10.55}$ × 1 = 0.095 m^3

Volume of fine aggregates = $\frac{1}{10.55}$ × 3 = 0.284 m^3

Volume of coarse aggregates = $\frac{1}{10.55}$ × 6 = 0.569 m^3

Volume of water = $\frac{1}{10.55}$ × 0.55 = 0.052 m^3 or 52 litres

The mass of each ingredient can be calculated by multiplying its volume by density:

Mass of 0.095 m^3 of cement = 0.095 × 1506 = 143.1 kg

Mass of 0.284 m^3 of sand = 0.284 × 1650 = 468.6 kg

Mass of 0.569 m^3 of gravel = 0.569 × 1600 = 910.4 kg

Mass of 0.052 m^3 of water = 0.052 × 1000 = 52 kg

Total = 1574 kg/m^3

The density of fresh concrete is 2350 kg/m^3 (approx.). The ratio in which the volumes of the dry materials should be increased is $\frac{2350}{1574}$ or 1.49 (say 1.5). The volumes of dry materials that will give 1 m^3 of concrete after mixing are:

Volume of cement = 0.095 × 1.5 = 0.143 m^3

Volume of sand = 0.284 × 1.5 = 0.426 m^3

Volume of gravel = 0.569 × 1.5 = 0.854 m^3

Volume of water = 0.052 × 1.5 = 0.078 m^3 or 78 litres

APPENDIX **2**

Solutions

Exercise 1.1

1. | 3 | 7 | . | 8 | 5 | − | 4 | 0 | . | 6 | 2 | + |

   | 3 | 1 | . | 8 | 5 | − | 9 | . | 6 | 7 | = | **19.41**

2. | 3 | 3 | . | 9 | × | 5 | 6 | . | 3 | ÷ | 4 | 5 | . |

   | 6 | 6 | = | **41.8**

3. | 6 | 7 | . | 3 | × | 6 | 9 | . | 8 | 1 | ÷ | 2 | 5 |

   | . | 9 | 7 | ÷ | 2 | 0 | . | 5 | = | **8.83**

4. | √ | 4 | . | 9 | × | √ | 8 | . | 5 | + | √ | 7 |

   | . | 4 | = | **9.17**

5. | π | × | 1 | 2 | . | 2 | 5 | x^2 | = | **471.44**

6. (a) | (| 2 | . | 2 | × | 9 | . | 8 |) | + | (| 5 |

   | . | 2 | × | 6 | . | 3 |) | = | **54.32**

   (b) | (| 4 | . | 6 | 6 | × | 1 | 2 | . | 8 |) | - |

   | (| 7 | . | 5 | × | 5 | . | 9 | 5 |) | = | **15.02**

   (c) | (| 4 | . | 6 | × | 1 | 0 | . | 8 |) | ÷ | (|

   | 7 | . | 3 | × | 5 | . | 5 |) | = | **1.24**

7. (a) `5` `^` `3` `×` `3` `^` `4` `÷` `2` `^` `5` `=` **316.41**

 (b) `4` `^` `3` `×` `6` `^` `3` `÷` `5` `^` `4` `=` **22.12**

8. `1` `0` `Log` `(` `9` `EXP` `+/−` `8` `÷` `2` `EXP` `+/−` `1` `1` `)` `=` **36.53**

9. (a) `sin` `7` `0` `÷` `cos` `6` `0` `=` **1.879**

 (b) `tan` `4` `5` `÷` `cos` `3` `5` `=` **1.221**

10. (a) `SHIFT` `sin` `0` `.` `8` `5` `=` 58.2117° `°′″` **58°12′42″**

 (b) `SHIFT` `cos` `0` `.` `7` `5` `=` 41.4096° `°′″` **41°24′34.64″**

 (c) `SHIFT` `tan` `0` `.` `6` `6` `=` 33.4248° `°′″` **33°25′29.32″**

11. (a) `sin` `6` `2` `°′″` `4` `2` `°′″` `3` `5` `°′″` `=` **0.8887**

 (b) `cos` `3` `2` `°′″` `2` `2` `°′″` `3` `5` `°′″` `=` **0.8445**

 (c) `tan` `8` `5` `°′″` `1` `0` `°′″` `2` `0` `°′″` `=` **11.8398**

Exercise 2.1

1. $34.21 + 26.05 + 370.30 =$ **430.56**
2. $34.11 - 20.78 =$ **13.33**
3. $309.1 - 206.99 - 57.78 =$ **44.33**
4. $40.0 \times 0.25 =$ **10.0**
5. $25 \times 125 \div 625 =$ **5**
6. (a) $379 \times 100 =$ **37 900**
 (b) $39.65 \times 1000 =$ **39 650**
 (c) $39.65 \times 10\,000 =$ **396 500**
7. (a) $584 \div 100 =$ **5.84**
 (b) $45.63 \div 100 =$ **0.4563**
 (c) $45.63 \div 1000 =$ **0.04563**
8. (a) $6 - 2 - 3(4 \times 1.5 \times 0.5) + 15$
 $= 6 - 2 - 3(3) + 15 =$ **10**
 (b) $15 - (2 \times 4 + 2) - (4 - 3)$
 $= 15 - (10) - 1 =$ **4**
9. $- 2° - 5° =$ **− 7°C**

10. (a) £29.85 + £19.98 + £15.50 = **£65.33**

(b) (20 × 4) − 65.33 = **£14.67**

11. (a) Length of skirting board required = 4.2 + 4.2 + 3.8 + 3.8 − 0.86

= 15.14 m

Lengths of 2.4 m long skirting board required = 15.14 ÷ 2.4 = 6.31

Two packs will be required; cost = 2 × £34.99 = **£69.98**

(b) Length of coving = 4.2 + 4.2 + 3.8 + 3.8 = 16 m

Lengths required (3 m long) = 16 ÷ 3 = 5.33, say 6

Cost = £6.00 × 6 = **£36.00**

12. Height of the walls = 2.8 − 0.1 − 0.1 = 2.6 m

Wall area (gross) = 2.6 × (4.2 + 3.8 + 4.2 + 3.8) = 41.6 m²

Wall area (net) = 41.6 − 2.06 × 0.86 − 2.0 × 1.2 = 37.43 m²

Area of one roll = 10.0 × 0.52 = 5.2 m²

Number of rolls required = 37.43 ÷ 5.2 = 7.2

Number of rolls, including wastage = $7.2 + 7.2 \times \dfrac{15}{100}$ = 8.28 or 9 rolls

Cost = 9 × £21.99 + £25.00 = **£222.91**

Exercise 3.1

1. (a) $5a + 2b + 3b - 2a = 5a - 2a + 2b + 3b = \mathbf{3a + 5b}$

(b) $5x - 3y - 2x - 3y = 5x - 2x - 3y - 3y = \mathbf{3x - 6y}$

2. $2xy^3 \times 5x^3y^3 = 2 \times 5 \times (x \times x^3) \times (y^3 \times y^3) = \mathbf{10x^4y^6}$

3. $\dfrac{25a^3b^2c^4}{5a^3bc^3} = 5a^{3-3}b^{2-1}c^{4-3} = 5a^0b^1c^1 = \mathbf{5bc}$

4. (a) $4(2x + 3y) = 4 \times 2x + 4 \times 3y = \mathbf{8x + 12y}$

(b) $2(3x - 6y) = \mathbf{6x - 12y}$

(c) $5 + (x + 2y + 10) = \mathbf{x + 2y + 15}$

(d) $3 - (-x + 3y - 4z) = \mathbf{3 + x - 3y + 4z}$

5. (a) $x - 5 = 14; x = 14 + 5$ or $\mathbf{x = 19}$

(b) $2a = 15; a = \dfrac{15}{2} = \mathbf{7.5}$

(c) $\dfrac{y}{5} = 3.5; y = 3.5 \times 5 = \mathbf{17.5}$

(d) $3(2x + 4) - 2(x - 3) = 4(2x + 4)$

$6x + 12 - 2x + 6 = 8x + 16$

$12 + 6 - 16 = 8x - 6x + 2x$

$4x = 2$; or $x = \dfrac{2}{4} = \mathbf{0.5}$

(e) $\dfrac{x}{2} + \dfrac{x+3}{4} = \dfrac{2x+1}{3}$

Multiply both sides by 12: $6x + 3(x + 3) = 4(2x + 1)$

$6x + 3x + 9 = 8x + 4; 9x - 8x = 4 - 9;$ or $\mathbf{x = -5}$

6. Let $\angle B = x$; $\therefore \angle A = 2x$

   $x + 2x + 75 = 180°$; $3x = 105°$; $\therefore \textbf{x = 35°}$

   $\angle B = x = 35°$; $\angle A = 2x = 70°$

7. Let the width be x cm; length = $2x$

   Perimeter = $x + 2x + x + 2x = 42$; therefore x = 7 cm

   width = 7 cm; length = 2×7 = **14 cm**

8. Let the width be x cm; length = $x + 3$ cm

   Perimeter = $x + (x + 3) + x + (x + 3) = 30$

   $4x + 6 = 30$; therefore $x = 6$ cm

   Width = 6 cm; Length = $6 + 3$ = **9 cm**

9. $2x + x + (x + 40) + (x + 20) = 360$

   $5x + 60 = 360$; or $x = 60°$

   $\angle A = 120°$, $\angle B = 60°$, $\angle C = 60° + 40° = 100°$, $\angle D = 60° + 20° = 80°$

10. Price of plot B = x (£); Price of plot A = $x - 20\ 000$

    Price of plot C = $x + 30\ 000$

    $x + (x - 20\ 000) + (x + 30\ 000) = £190\ 000$

    $3x + 10\ 000 = 190\ 000$; or $x = 60\ 000$

    Prices: **Plot B = £60 000; Plot A = £40 000; Plot C = £90 000**

11. $8 + 12 + x - 6 + 12 - x + x - 4 + x = 40$

    $22 + 2x = 40$; $2x = 18$; or $\textbf{x = 9 m = DE}$

    $AB = x - 6 = 3$ m; $BC = 12 - x = 3$ m; $CD = x - 4 = 5$ m

Exercise 4.1

1. (a) $4^2 \times 4^7 = 4^{2+7} = 4^9 = \textbf{262 144}$

   (b) $m^2 \times m^3 = m^{2+3} = \textbf{m}^5$

2. (a) $3^2 \times 3^5 \times 3^7 = 3^{2+5+7} = 3^{14} = \textbf{4 782 969}$

   (b) $4 \times 4^2 \times 4^7 = 4^{1+2+7} = 4^{10} = \textbf{1 048 576}$

3. (a) $\dfrac{n^3 \times n^4}{n^1 \times n^5} = n^{7-6} = \textbf{n}^1 = \textbf{n}$

   (b) $\dfrac{x^6 \times x^2}{x^3 \times x} = x^{8-4} = \textbf{x}^4$

4. (a) $(2^3)^5 = 2^{3 \times 5} = 2^{15} = \textbf{32 768}$

   (b) $(3x^2)^4 = 3^4 x^{2 \times 4} = \textbf{81x}^8$

5. $\dfrac{3^4}{3^7} = 3^{4-7} = 3^{-3} = \dfrac{1}{3^3}$ or $\dfrac{1}{27}$

6. (a) $\dfrac{a^3}{a^2 \times a^5} = a^{3-7} = \textbf{a}^{-4}$

   (b) $\dfrac{x^4 \times x^2}{x^3 \times x^5} = x^{6-8} = \textbf{x}^{-2}$

7. (a) $\dfrac{a^1 \times a^2 \times a^4}{a^3 \times a^4} = a^{7-7} = a^0 = \mathbf{1}$

(b) $\dfrac{y^6 \times y^4}{y^3 \times y^5 \times y^2} = y^{10-10} = y^0 = \mathbf{1}$

8. (a) | log | 2 | 5 | = | **1.3979**

(b) | log | 1 | 5 | 0 | = | **2.1761**

(c) | log | 1 | 2 | 0 | 4 | = | **3.0806**

9. (a) | log | 2 | . | 2 | EXP | 2 | = | **2.3424**

(b) | log | 3 | . | 8 | EXP | - | 3 | = | **−2.4202**

10. (a) | SHIFT | LOG | 8 | . | 5 | = | **316 227 766**

(b) | SHIFT | LOG | 0 | . | 7 | 2 | = | **5.248**

(c) | SHIFT | LOG | 0 | . | 0 | 0 | 1 | 4 | = | **1.003**

Exercise 5.1

1. (a) $976 = \mathbf{9.76 \times 10^2}$
   (b) $1478 = \mathbf{1.478 \times 10^3}$
   (c) $377\,620 = \mathbf{3.7762 \times 10^5}$
2. (a) $0.025 = \mathbf{2.5 \times 10^{-2}}$
   (b) $0.00071 = \mathbf{7.1 \times 10^{-4}}$
   (c) $0.000000437 = \mathbf{4.37 \times 10^{-7}}$
3. (a) $1.721 \times 10^2 = \mathbf{172.1}$
   (b) $2.371 \times 10^{-3} = \mathbf{0.002371}$
   (c) $9.877 \times 10^4 = \mathbf{98770}$
   (d) $9.1 \times 10^{-6} = \mathbf{0.0000091}$
4. **361.73**   correct to 5 s.f.
   361.7   correct to 4 s.f.
   362   correct to 3 s.f.
   360   correct to 2 s.f.
5. **867 360**   correct to 5 s.f.
   867 400   correct to 4 s.f.
   867 000   correct to 3 s.f.
   870 000   correct to 2 s.f.
   900 000   correct to 1 s.f.

6. **0.00084**   correct to 2 s.f.

   0.0008   correct to 1 s.f.

7. Estimated answer   Actual answer

   (a) $450 + 150 + 750 = $ **1350**   1352

   (b) $40 \times 15 = $ **600**   570

   (c) $\dfrac{66 \times 90}{11 \times 30} = $ **18**   **17.019**

Exercise 6.1

1. $b = c - d - a$

   After transposition, $\boldsymbol{a = c - d - b}$

2. (a) $y = mx + c$; this can be written as $mx + c = y$

   After transposition, $\boldsymbol{c = y - mx}$

   (b) $c = 10 - 4 \times 2 = $ **2**

3. $v = u + at$, which can be written as $u + at = v$

   $at = v - u$ or $\boldsymbol{t = \dfrac{v - u}{a}}$

4. $p = \dfrac{\pi d}{2}$ or $\dfrac{\pi d}{2} = p$

   $\pi d = 2p$ or $\boldsymbol{d = \dfrac{2p}{\pi}}$

5. (a) $\dfrac{k A \theta}{d} = Q$; after transposition, $kA\theta = Qd$

     Transpose $A\theta$ to the RHS, $\boldsymbol{k = \dfrac{Qd}{A\theta}}$

   (b) $k = \dfrac{2000 \times 0.5}{10 \times 40} = $ **2.5**

6. (a) $A = 4\pi r^2$ or $4\pi r^2 = A$

     Transpose 4π to the RHS, $r^2 = \dfrac{A}{4\pi}$

     $\therefore \boldsymbol{r = \sqrt{\dfrac{A}{4\pi}}}$

   (b) $r = \sqrt{\dfrac{5000}{4\pi}} = $ **19.95 cm**

7. $V = \pi r^2 h$ or $\pi r^2 h = V$

   (a) Transpose πr^2 to the RHS, $\boldsymbol{h = \dfrac{V}{\pi r^2}}$

   (b) $r^2 = \dfrac{V}{\pi h}$ or $\boldsymbol{r = \sqrt{\dfrac{V}{\pi h}}}$

8. (a) $2as = v^2 - u^2$ or $s = \dfrac{v^2 - u^2}{2a}$

(b) $s = \dfrac{5^2 - 0^2}{2 \times 2.5} = \mathbf{5\ m}$

9. $mx + c = y$

After transposition, $mx = y - c$ or $x = \dfrac{y - c}{m}$

10. $y = \dfrac{3x + 2z}{7} + d$

Multiply both sides by 7:

$7y = 3x + 2z + 7d$

$3x = 7y - 2z - 7d$

or $x = \dfrac{7y - 2z - 7d}{3}$

11. $V = \dfrac{\pi}{3}\, r^2 h$

$\pi r^2 h = 3V$

$r^2 = \dfrac{3V}{\pi h}$

$\therefore\ \mathbf{r = \sqrt{\dfrac{3V}{\pi h}}}$

12. (a) $v = c\sqrt{mi}$, which can be written as $c\sqrt{mi} = v$

Square both sides, $c^2 mi = v^2$ or $\mathbf{m = \dfrac{v^2}{c^2 i}}$

(b) $m = \dfrac{(2.42)^2}{50^2 \times 0.025} = \mathbf{0.0937}$

13. $\dfrac{1}{R} = \dfrac{1}{R_1} + \dfrac{1}{R_2} = \dfrac{1}{5} + \dfrac{1}{10} = 0.2 + 0.1 = 0.3$

$\dfrac{1}{R} = 0.3,$ or $R = \dfrac{1}{0.3} = \mathbf{3.33\ \Omega}$

14. Rate of heat loss due to ventilation $= \dfrac{C_v \times V \times N \times T}{3600}$

$C_v = 1212$ J/m^3 °C, $V = 400$ m^3, $N = 2.0$ and $T = 21$°C

Rate of heat loss due to ventilation $= \dfrac{1212 \times 400 \times 2 \times 21}{3600} = \mathbf{5656\ W}$

15. One litre of water has a mass of 1 kg.

Heat energy required $= M \times S \times T$ ($T = 50 - 10 = 40$°C)

$= 162 \times 4.186 \times 40 = \mathbf{27\,125.3\ kJ}$

Assuming no loss of heat energy, power required to heat the water in

one hour $= \dfrac{27\,125.3}{3600} = \mathbf{7.53\ kW}$

16. (a) Water pressure at the base of the vessel,

$p = \rho gh$

$= 1000 \times 9.81 \times 4$

$= \textbf{39 240 N/m}^2 \textbf{ or 39 240 Pa}$

Note: 1 Newton/m^2 = 1 Pascal (Pa)

(b) Force = pressure × area

$= 39\,240 \times (1.5 \times 1.2)$

$= \textbf{70 632 N or 70.632 kN}$

17. (a) Radius of the drain = 75 mm or 0.075 m

Cross-sectional area of flow = $\dfrac{\pi \times (0.075)^2}{2} = 0.008836$ (the drain runs

half-full)

Slope, S = 1 in 80 or $\dfrac{1}{80} = 0.0125$

Velocity of flow, $v = c\sqrt{RS} = 50\sqrt{0.075 \times 0.0125}$

$= 50 \times 0.0306 = \textbf{1.53 m/s}$

(b) Flow rate = $A \times v = 0.008836 \times 1.53 = \textbf{0.0135 m}^3\textbf{/s}$

Exercise 7.1

1. (a) $\dfrac{3}{9} = \dfrac{1}{3}$

   (b) $\dfrac{3}{9} = \dfrac{1}{3}$

2. (a) $\dfrac{20}{120} = \dfrac{1}{6}$

   (b) $\dfrac{250}{1200} = \dfrac{25}{120} = \dfrac{5}{24}$

   (c) $\dfrac{60}{120} = \dfrac{1}{2}$

3. (a) $\dfrac{10}{12} = \dfrac{5}{6}$

   (b) $\dfrac{12}{15} = \dfrac{4}{5}$

   (c) $\dfrac{40}{100} = \dfrac{2}{5}$

4. (a) $\dfrac{2}{3} + \dfrac{5}{6} = \dfrac{2\times 2 + 1\times 5}{6} = \dfrac{9}{6} = \dfrac{3}{2}$

   (b) $\dfrac{1}{3} + \dfrac{1}{5} = \dfrac{1\times 5 + 1\times 3}{15} = \dfrac{8}{15}$

(c) $\dfrac{3}{4} + \dfrac{2}{5} = \dfrac{3\times5 + 2\times4}{20} = \mathbf{\dfrac{23}{20}}$

(d) $\dfrac{3}{4} + \dfrac{4}{5} = \dfrac{3\times5 + 4\times4}{20} = \mathbf{\dfrac{31}{20}}$

5. (a) $\dfrac{1}{3}$ of $33 = \dfrac{1}{3} \times 33 = \mathbf{11\ m^2}$

(b) $\dfrac{1}{5}$ of $1.50 = \dfrac{1}{5} \times 1.50 = \mathbf{£0.30}$

(c) $\dfrac{3}{4}$ of $200 = \dfrac{3}{4} \times 200 = \mathbf{150\ m}$

6. (a) $\dfrac{3}{4} = \mathbf{0.75 = 75\%}$

(b) $\dfrac{3}{5} = \mathbf{0.60 = 60\%}$

(c) $\dfrac{7}{10} = \mathbf{0.70 = 70\%}$

(d) $\dfrac{4}{5} = \mathbf{0.80 = 80\%}$

7. (a) $\dfrac{20}{100} \times 150 = \mathbf{30}$

(b) $\dfrac{75}{100} \times 3000 = \mathbf{2250}$

(c) $\dfrac{90}{100} \times 900 = \mathbf{810}$

8. Maths: $\dfrac{40}{60} \times 100 = 66.67\%$; science: $\dfrac{50}{80} \times 100 = 62.5\%$

Nikki did better in **Maths**.

9. $\dfrac{25}{100} \times 200 = £50$ (Peter); $\dfrac{30}{100} \times 200 = £60$ (Karen); $\dfrac{15}{100} \times 200 = £30$

(Dave); Money left $= 200 - 50 - 60 - 30 = \mathbf{£60}$

10. $\dfrac{4.5}{100} \times 1200 = \mathbf{£54}$

11. Total without VAT $= 845.00 + (7 \times 80) = 1405.00$

$\text{VAT} = \dfrac{20}{100} \times 1405.00 = 281.00$

Total (including VAT) $= 1405.00 + 281.00 = \mathbf{£1686.00}$

12. Total mass of concrete $= 1500$ kg

(a) $\dfrac{450}{1500} = \dfrac{30}{100} = \mathbf{\dfrac{3}{10}}$ **or 30%**

(b) $\dfrac{100}{200} = \mathbf{\dfrac{1}{2}}$ **or 50%**

13. Brick A: $\dfrac{0.2}{2.200} = \dfrac{2}{22} = \dfrac{1}{11}$ **or 9.09%**

Brick B: $\dfrac{0.25}{2.5} = \dfrac{2.5}{25} = \dfrac{1}{10}$ **or 10%**

14. Heat loss through windows $= \dfrac{15}{100} \times 25\,000 =$ **3750 W**

Heat loss through roof $= \dfrac{35}{100} \times 25\,000 =$ **8750 W**

15. $\dfrac{27}{216} = \dfrac{3}{24} = \dfrac{1}{8}; \dfrac{1}{8} \times 100 =$ **12.5%**

16. Total money spent = £6000

(a) Money spent on foundation and brickwork $= \dfrac{2200}{6000} = \dfrac{22}{60} = \dfrac{11}{30}$

(b) Money spent on finishes $= \dfrac{1000}{6000} \times 100 =$ **16.67%**

17. Moisture content $= \dfrac{(2.35 - 2.0)}{2.0} \times 100 =$ **17.5%**

Exercise 8.1

1.

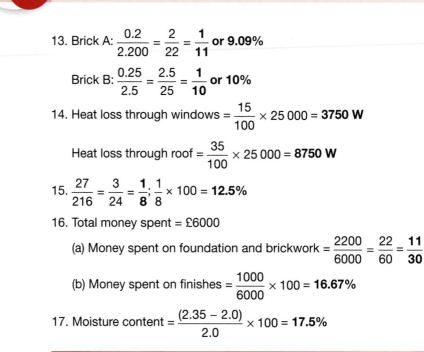

2.

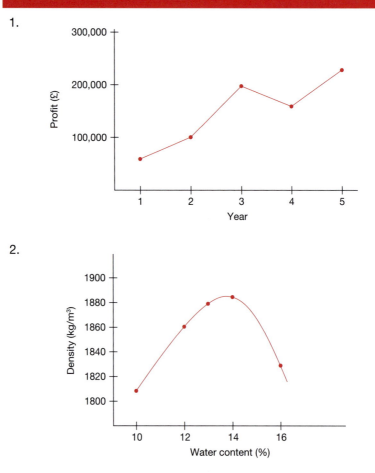

3.

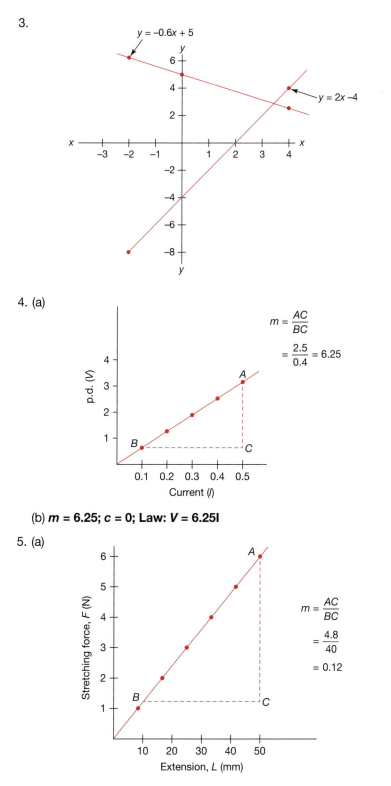

$y = -0.6x + 5$

$y = 2x - 4$

4. (a)

$m = \dfrac{AC}{BC}$

$= \dfrac{2.5}{0.4} = 6.25$

p.d. (V)

Current (I)

(b) **m = 6.25; c = 0; Law: V = 6.25I**

5. (a)

Stretching force, F (N)

$m = \dfrac{AC}{BC}$

$= \dfrac{4.8}{40}$

$= 0.12$

Extension, L (mm)

(b) **m = 0.12; c = 0; Law: F = 0.12L**

6. (a)

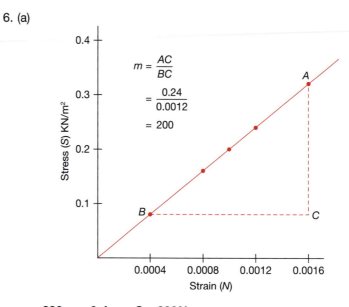

$$m = \frac{AC}{BC}$$

$$= \frac{0.24}{0.0012}$$

$$= 200$$

m = 200; c = 0; Law: S = 200N

(b) **S = 0.4 kN/m²**

7. (a) Consider 20°C as the initial temperature

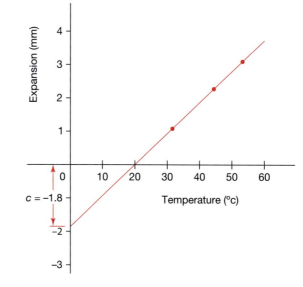

m = 0.091; c = −1.8; Law: E = 0.091T − 1.8

(b) **999.11 mm**

8. **m = −4.583; c = 45.67; Law: Y = −4.583x + 45.67**

9.

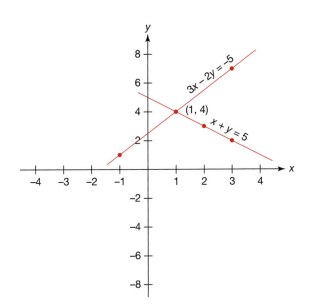

x = 1, y = 4

10.

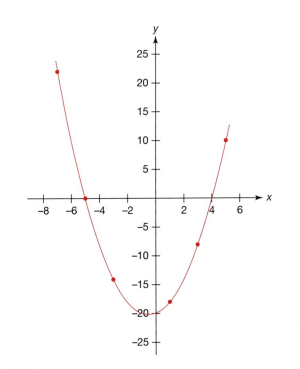

x = −5, x = 4

Exercise 9.1

1. (a) $86 \text{ mm} = \dfrac{86}{1000} = \textbf{0.086 m}$

   (b) $385 \text{ mm} = \dfrac{385}{1000} = \textbf{0.385 m}$

   (c) $7.4 \text{ m} = 7.4 \times 100 = \textbf{740 cm}$

   (d) $3 \text{ ft } 2 \text{ in} = (38 \text{ in}) \times 25.4 = \textbf{965.2 mm}$

   (e) $1 \text{ yd} = 0.9144 \text{ m}; \; 6 \text{ yd} = 6 \times 0.9144 = \textbf{5.4864 m}$

   (f) $14 \text{ in} = 14 \times 25.4 = 355.6 \text{ mm, or } \textbf{0.3556 m}$

2. (a) $24 \text{ ft} = 24 \times 0.3048 = \textbf{7.315 m}$

   (b) $9 \text{ m} = 9/0.3048 = 29.528 \text{ ft} = 29 \text{ ft} + (0.528 \times 12 \text{ in}) = \textbf{29 ft 6.3 in}$

3. (a) $0.070 \text{ kg} \times 1000 = \textbf{70 g}$

   (b) $25\,500 \text{ kg} = \dfrac{25\,500}{1000} = \textbf{25.5 tonnes}$

   (c) $815 \text{ g} = 815 \times 0.03527 = \textbf{28.745 oz}$

4. (a) $15.500 \text{ kg} = 15.5 \times 2.20462 = \textbf{34.172 lb}$

   (b) $250 \text{ kg} = 250 \times 2.20462 = \textbf{551.155 lb}$

   (c) $75 \text{ lb} = \dfrac{75}{2.20462} = \textbf{34.019 kg}$

5. (a) $12\,550 \text{ mm}^2 = 12\,550 \times 0.01 = \textbf{125.5 cm}^2$

   (b) $655\,000 \text{ mm}^2 = 655\,000 \times 10^{-6} = \textbf{0.655 m}^2$

   (c) $2.75 \text{ m}^2 = 2.75 \times 10^6 = \textbf{2\,750\,000 mm}^2$

   (d) $5.25 \text{ m}^2 = 5.25 \times 10\,000 = \textbf{52\,500 cm}^2$

6. (a) $95\,450 \text{ mm}^3 = 95\,450 \times 10^{-9} = \textbf{9.545} \times \textbf{10}^{\textbf{-5}} \textbf{ m}^3$

   (b) $9545 \text{ mm}^3 = 9545 \times 0.001 = \textbf{9.545 cm}^3$

   (c) $0.00094 \text{ m}^3 = 0.00094 \times 10^9 = \textbf{940\,000 mm}^3$

7. (a) $12\,560 \text{ ml} = 12\,560 \times 0.001 = \textbf{12.560 litres}$

   (b) $6.67 \text{ litres} = 6.67 \times 100 = \textbf{667 centilitres}$

   (c) $350 \text{ centilitres} = 350 \times 0.01 = \textbf{3.50 litres}$

   (d) $0.0053 \text{ m}^3 = 0.0053 \times 1000 = \textbf{5.3 litres}$

   (e) $25 \text{ litres} = 25 \times 0.22 = \textbf{5.5 gallons}$

8. $37°C = 95 + 3.6 = \textbf{98.6°F}$

   $61.5°C = 140 + 2.7 = \textbf{142.7°F}$

Exercise 10.1

1. (a) $72°38'51'' = 72° + \left(\dfrac{38}{60}\right) + \left(\dfrac{51}{60 \times 60}\right) = \textbf{72.6475°}$

   (b) $63.88° = 63° + \left(0.88 \times 60\right)' = 63° + 52.8' = 63°52' + (0.8 \times 60)''$
   $= \textbf{63°52'48''}$

(c) $30°48'50'' = 30.81389° = (30.81389 \times \dfrac{2\pi}{360})$ radians

$\qquad\qquad = \textbf{0.5378 radians}$.

2. $\angle y = \textbf{107°}$ (opposite angles); $\angle y + \angle z = 180°$

$\quad \angle z = 180° - \angle y = 180° - 107° = \textbf{73°}$

$\quad \angle x = 360° - 76° - 73° - 107° = \textbf{104°}$

3. (a) Unknown side $= \sqrt{7^2 + 6^2} = \textbf{9.220 m}$

(b) Unknown side $= \sqrt{10.5^2 - 5.5^2} = \textbf{8.944 m}$

(c) Unknown side $= \sqrt{11.8^2 - 9.4^2} = \textbf{7.133 m}$

4. (a) Line XY is parallel to line PN, therefore $\angle MPN$ and $\angle MQR$ are equal. $\angle PMN$ is common to both triangles, therefore the triangles are similar.

(b) $\dfrac{40}{80} = \dfrac{PM}{110}$ $\therefore PM = \dfrac{40 \times 110}{80} = 55\text{m}$

$\quad PQ = 55 - 40 = \textbf{15 m}$

$\quad \dfrac{70}{MN} = \dfrac{80}{110}$ $\therefore MN = \dfrac{70 \times 110}{80} = 96.25 \text{ m}$

$\quad RN = MN - MR = 96.25 - 70 = \textbf{26.25 m}$

5. All angles of $\triangle BCF$ are equal (60°) therefore all sides are equal as well. In $\triangle BAF$ side BA must be equal to side BF. $\therefore BC = CF = BF = BA = \textbf{3 m}$

$\quad AF = FE = \sqrt{6^2 - 3^2} = \textbf{5.196 m}$

6. Length of third boundary $= \sqrt{38^2 + 26^2} = \textbf{46.043 m}$

7. $\angle B = \angle D = \textbf{113°}$

$\quad \angle A = \angle C = \tfrac{1}{2}(360 - 2 \times 113) = \textbf{67°}$

8. Circumference $= 2\pi r = 2 \times \pi \times 5.6 = \textbf{35.186 m}$

Exercise 11.1

1. (a) Area $= \tfrac{1}{2} \times 6 \times 8 = \textbf{24 cm}^2$

(b) $GD = \sqrt{12^2 - 4^2} = 11.3137$ cm

$\qquad \therefore$ area $= \tfrac{1}{2}(11.3137 \times 8) = \textbf{45.255 cm}^2$

(c) Area $= 11 \times 7.5 = \textbf{82.5 cm}^2$

(d) Area $= 7 \times 7 = \textbf{49 cm}^2$

(e) Area $= \tfrac{1}{2}(8 + 12) \times 6 = \textbf{60 cm}^2$

(f) Area $= 11 \times 5.5 = \textbf{60.5 cm}^2$

2. (a) Area: circle $= \pi \times 1.75^2 = \textbf{9.621 m}^2$

(b) Area: major sector $= \dfrac{245}{360} \times 9.621 = \textbf{6.548 m}^2$

$\qquad$ Area: minor sector $= \dfrac{115}{369} \times 9.621 = \textbf{3.073 m}^2$

3. Area of concrete path $= \pi \times 16^2 - \pi \times 15^2 = \textbf{97.389 m}^2$

4. Length (including path) = 15 + 1.2 + 1.2 = 17.4 m

   Width (including path) = 11 + 1.2 + 1.2 = 13.4 m

   Area = (17.4 × 13.4) − (15 × 11) = **68.16 m²**

5. Area of girder = (12 × 1.5) + (20 × 1.5) + (10 × 1.0) = **58 cm²,** or **5800 mm²**

6. Area of floor = 3.50 × 3.00 = 10.5 m²

   Area of one tile = 0.33 × 0.33 = 0.1089 m²

   Therefore number of tiles = $\dfrac{10.5}{0.1089}$ = 96.42

   Add 10% = 1.1 × 96.42 = 106.06 tiles

   Number of packs required = $\dfrac{106.06}{9}$ = 11.78, so **12 packs**

7. Area of drive = 9.0 × 8.0 $- \dfrac{1}{4}\pi \times 4^2$ = 59.434 m²

   Area of one block = 0.2 × 0.1 = 0.02 m²

   Number of blocks = $\dfrac{59.434}{0.02}$ = 2971.68

   Add 10% for cutting/wastage = 1.1 × 2971.68 = 3268.9, so **3269 blocks**.

8. Area of floor = 5.0 × 4.0 = 20.0 m²

   Number of packs = $\dfrac{20.0}{2.106}$ = 9.5

   Add 10% for wastage = 1.1 × 9.5 = 10.45 packs, so **11 packs**.

9. Length of internal walls = 2(4.8 + 3.8) = 17.2 m

   Height of internal walls = 2.8 − 0.100 = 2.7 m

   Wall area (including door and window) = 17.2 × 2.7 = 46.44 m²

   Area of door and window = 1.9 × 1.0 + 1.8 × 1.2 = 4.06 m²

   Net wall area = 46.44 − 4.06 = **42.38 m²**

   Area of ceiling = 4.8 × 3.8 = **18.24 m²**

10. Area of one roll = 0.52 × 10.0 = 5.2

    Number of rolls = $\dfrac{\text{wall area}}{5.2} = \dfrac{42.38}{5.2}$ = 8.15

    Add 15% = 1.15 × 8.15 = 9.37, so **10 rolls**.

11. Area of the wall = 5.0 × 2.4 = 12.0 m²

    Area of one brick = 0.016875 m²; number of bricks

    $$= \dfrac{12.0}{0.016\ 875} = 711.1$$

    Add 5% for wastage = 1.05 × 711.1 = 746.7, so **747 bricks**

    Area of one block = 0.101250 m²; number of blocks

    $$= \dfrac{12.0}{0.101\ 250} = 118.5$$

    Add 5% = 1.05 × 118.5 = 124.4, so **125 blocks**

12. Area of gable end $= 12 \times 5.4 + \dfrac{12 \times 4.0}{2} = 88.8$ m^2

Number of bricks $= \dfrac{88.8}{0.016\,875} = 5262.2$ bricks

Add 5% = 5262.2 × 1.05 = 5525.3, so **5526 bricks**

Number of blocks $= \dfrac{88.8}{0.101\,250} = 877.04$ blocks

Add 5% = 877.03 × 1.05 = 920.89, so **921 blocks**

Exercise 12.1

1. Volume = 3.24 × 4.38 × 2.46 = **34.91 m^3**

2. Radius = 225 mm or 0.225 m; volume = $\pi \times (0.225)^2 \times 0.950$

$\qquad\qquad\qquad\qquad\qquad\qquad = $ **0.15109 m^3**

0.15109 m^3 = 0.151 09 × 1000 = **151.09 litres**

3. Volume $= \dfrac{1}{3} \times (3.5 \times 3.5) \times 6.0 = $ **24.5 m^3**

4. Radius of the base = 1.5 m

Volume $= \dfrac{1}{3} \times \pi \times (1.5)^2 \times 8 = $ **18.85 m^3**

5. Volume $= (300 \times 300 \times 400) + \left[\dfrac{1}{3} \times (300 \times 300) \times 400 \right]$

$\qquad = $ **48 000 000 mm^3** or **0.048 m^3**

6. Volume $= \left(\dfrac{1}{2} \times 12 \times 4 \right) \times 14 = $ **336 m^3**

7. Internal size of the culvert is 1.45 m × 0.95 m

Cross-sectional area of culvert = 1.75 × 1.25 − 1.45 × 0.95 = 0.81 m^2

Volume of concrete per linear metre = 0.81 × 1.0 = **0.81 m^3**

8. Cross-sectional area $= \dfrac{1}{2} \times (10.85 + 8.40) \times 2.95 = 28.39375$ m^2

Volume = 28.39375 × 25 = **709.844 m^3**

9. (a) Volume of water = 2.4 × 1.8 × (1.20 − 0.15) = **4.536 m^3**

(b) Mass = Volume × Density = 4.536 × 1000 = **4536 kg**

(c) Force on the joists = 4536 × 9.8 = **44 452.8 N**

10. (a) Volume = 8.5 × 3.5 × 0.150 = **4.4625 m^3**

(b) Quantities: Cement = 0.198 × 4.4625 = **0.884 m^3**

$\qquad\qquad\qquad$ Sand = 0.398 × 4.4625 = **1.772 m^3**

$\qquad\qquad\qquad$ Gravel = 0.8 × 1.772 = **3.57 m^3**

$\qquad\qquad\qquad$ Water = 109 × 4.4625 = **486 litres**

11. (a) Plan area of foundation trench: Area 1 = 4.754 × 0.6 = 2.8524 m^2

$\qquad\qquad\qquad\qquad\qquad\qquad$ Area 2 = 2.651 × 0.6 = 1.5906 m^2

$\qquad\qquad\qquad\qquad\qquad\qquad$ Area 3 = 2.651 × 0.6 = 1.5906 m^2

Total plan area = 6.0336 m²; Volume of soil = 6.0336 × 0.9

$$= \textbf{5.430 m}^3$$

(b) Volume of soil after bulking $= \left(\dfrac{15}{100} \times 5.430\right) + 5.430 = \textbf{6.245 m}^3$

(c) Volume of concrete = 6.0336 × 0.2 = **1.207 m³**

Exercise 13.1

1. (a) sin 30°20′35″ = **0.5052**

    cos 50°10′30″ = **0.6404**

    tan 40°55′05″ = **0.8668**

   (b) sin⁻¹ 0.523 = **31°32′1.3″**

    cos⁻¹ 0.981 = **11°11′12.4″**

    tan⁻¹ 0.638 = **32°32′16.4″**

2. AC = 15 sin 30° = **7.5 cm**

   BC = 15 cos 30° = **12.99 cm**

3. $\angle BAC = 60°35′30″; \dfrac{BC}{AC} = \dfrac{\text{Opp}}{\text{Adj}} = \tan 60°35′30″$

   $BC = AC \times \tan 60°35′30″$

   $= 60 \times \tan 60°35′30″ = \textbf{106.447 m}$

4. In $\triangle ABC$, angle C is the right angle and AC represents the height of the building.

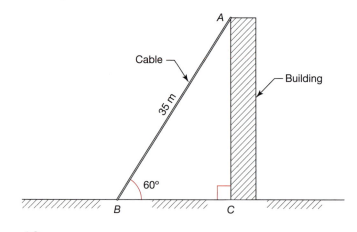

   (a) $\dfrac{AC}{AB} = \sin 60°; AC = AB \times \sin 60° = \textbf{30.311 m}$

   (b) $\dfrac{BC}{AB} = \cos 60°; BC = AB \times \cos 60° = \textbf{17.5 m}$

5. A gradient of 9% means that the road rises nine units for every 100 units of its horizontal measurement (see figure below). Let h = the rise or fall of the road.

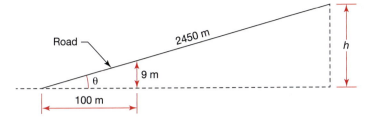

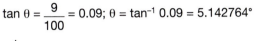

$\tan \theta = \dfrac{9}{100} = 0.09;\ \theta = \tan^{-1} 0.09 = 5.142764°$

$\dfrac{h}{2450} = \sin 5.142764°,\ \therefore h = 2450 \times \sin 5.142764° = \textbf{219.612 m}$

6. If θ is the pitch of the roof, then $\tan \theta = \dfrac{1}{48}$

$\therefore$ pitch $= \tan^{-1}\left(\dfrac{1}{48}\right) = 1.193°$ or **1°11′36.56″**

7.

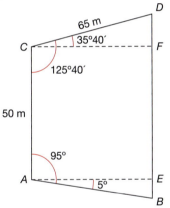

$CF = 65 \times \cos 35°40' = 52.8075$ m (also $CF = AE$)

$BD = DF + FE + EB = 65 \times \sin 35°40' + 50 + 52.8075 \times \tan 5°$

$= 92.519$ m

Area of plot (trapezium) $= \dfrac{50 + 92.519}{2} \times 52.8075 = \textbf{3763.04 m}^2$

8. (a) In ΔDEF, DF = height of roof = 5 tan 40° = 4.195 m

Common rafter $HJ = DE = \dfrac{DF}{\sin 40°} = \textbf{6.527 m}$

(b) In ΔDFA, $FA = \sqrt{5^2 + 5^2} = 7.071$ m

Hip rafter $DA = \sqrt{7.071^2 + 4.195^2} = \textbf{8.222 m}$

(c) $BD = \dfrac{DF}{\sin 40°} = 6.527$ m

Area of roof $= 2 \times [\dfrac{1}{2} \times (15 + 5) \times 6.527 + \dfrac{1}{2} \times 10 \times 6.527] = \textbf{195.81 m}^2$

9. Length of top of embankment = $40 - 2\left(\dfrac{3.0}{\tan 35°}\right) = 31.431$ m

   Cross-sectional area = $\dfrac{1}{2} \times (40 + 31.431) \times 3.0 =$ **107.147 m²**

   Volume = 1200×107.147 m² = **128 576.4 m³**

10. Pitch = $\tan^{-1}\left(\dfrac{220}{230}\right) =$ **43.73°**. Stairs are too steep.

    The pitch could be reduced by increasing the going and/or decreasing the rise.

    Floor to ceiling height = 220×11 risers = 2420 mm

    Try 12 risers, therefore new rise = $\dfrac{2420}{12} = 201.67$ mm

    New going = $\dfrac{201.67}{\tan 42°} = 223.98$ mm (say 224 mm)

    $2R + G = 2(201.67) + 224 = 627.34 < 700$; therefore satisfactory

    Use rise = 201.67 mm and going = 224 mm

Exercise 14.1

1. Refer to Figure 14.1 in Chapter 14.

   Stations A and B, and corner F form a triangle whose sides are: $AB = 116\ 861$ m; $AF = 74.447$ m; and $BF = 45\ 641$ m. The angles of $\triangle AFB$ may be calculated by sine/cosine rule which will enable the location of corner F of the factory unit.

   In $\triangle ABF$, $116.861^2 = 74.447^2 + 45.641^2 - 2(74.447)(45.641)\cos F$

   $\therefore \angle F =$ **152.559°**

   $\dfrac{116.861}{\sin 152.559} = \dfrac{45.641}{\sin A} \therefore \angle A =$ **10.369°**

   $\therefore \angle B = 180 - 152.558 - 10.369 =$ **17.072°**

   In $\triangle ACB$ distances AC and BC are 109.252 m and 11.722 m, respectively.

   $109.252^2 = 116.861^2 + 11.722^2 - 2(116.861)(11.722)\cos B$

   $\therefore \angle B =$ **47.301°**

   $\dfrac{109.252}{\sin 47.301} = \dfrac{116.861}{\sin C} = \dfrac{11.722}{\sin A} \therefore \angle A =$ **4.523°** and $\angle C =$ **128.176°**

   Having the necessary information the setting out will depend on the inter-visibility of the site stations and the equipment available. It is often done with a total station, but could be completed with a theodolite and tapes or EDM.

2. The angles at all corners must be 90°. The lengths of the diagonals must be:

   Diagonal $HB = \sqrt{7.56^2 + 7.0^2} =$ **10.30 m**

   Diagonal $AC = \sqrt{7.00^2 + (5.11 + 5.33 - 7.56)^2} =$ **7.57 m**

Diagonal $BD = \sqrt{6.55^2 + (5.11 + 5.33 - 7.56)^2} = \textbf{7.16 m}$

Diagonal $CE = \sqrt{5.33^2 + 6.55^2} = \textbf{8.44 m}$

Diagonal $EG = \sqrt{3.78^2 + 5.11^2} = \textbf{6.36 m}$

Diagonal $FH = \sqrt{5.11^2 + (7.00 + 6.55 + 3.78)^2} = \textbf{18.07 m}$

Diagonal $AG = \sqrt{7.56^2 + (7.0 + 6.55 + 3.78)^2} = \textbf{18.91 m}$

Diagonal $DH = \sqrt{(5.11 + 5.33)^2 + (7.0 + 6.55)^2} = \textbf{17.11 m}$

Diagonal $EH = \sqrt{5.11^2 + (7.0 + 6.55)^2} = \textbf{14.48 m}$

Diagonal $CG = \sqrt{(6.55 + 3.78)^2 + (5.11 + 5.33)^2} = \textbf{14.69 m}$

3. Refer to Figure 14.10. Take the right-hand quadrant of a 3.00 m diameter semicircle, divide it into six strips (i.e. seven ordinates) then calculate the lengths of the ordinates starting at the longest.

Ordinate number	Calculation	Length
1		**1.5 m**
2	$\sqrt{1.5^2 - .25^2}$	**1.479 m**
3	$\sqrt{1.5^2 - .5^2}$	**1.414 m**
4	$\sqrt{1.5^2 - .75^2}$	**1.299 m**
5	$\sqrt{1.5^2 - 1.0^2}$	**1.118 m**
6	$\sqrt{1.5^2 - 1.25^2}$	**0.829 m**
7	$\sqrt{1.5^2 - 1.5^2}$	**0.0 m**

The curve set out with these ordinates (each used twice: once for each quadrant) will describe a semicircle of 3.00 m diameter. A line parallel to the diameter must be drawn 100 mm from it to finish the setting out.

4. Eight strips will give nine ordinates.

$$R = \frac{\left(\dfrac{2.7}{2}\right)^2 + 1^2}{2 \times 1.0} = 1.411 \, \text{m}.$$ Now divide half the arch into eight strips and find the seven intermediate ordinates (the first and last are 1 m and 0.0 m). Let the lengths of ordinates be h_1, h_2, etc.

$h_1 = \textbf{1.000 m}$

$h_2 = \sqrt{1.411^2 - (1 \times 0.16875)^2} - 0.411 = \textbf{0.990 m}$

$h_3 = \sqrt{1.411^2 - (2 \times 0.16875)^2} - 0.411 = \textbf{0.959 m}$

$h_4 = \sqrt{1.411^2 - (3 \times 0.16875)^2} - 0.411 = \textbf{0.906 m}$

$h_5 = \sqrt{1.411^2 - (4 \times 0.16875)^2} - 0.411 = \textbf{0.828 m}$

$h_6 = \sqrt{1.411^2 - (5 \times 0.16875)^2} - 0.411 = \textbf{0.720 m}$

$h_7 = \sqrt{1.411^2 - (6 \times 0.16875)^2} - 0.411 = \textbf{0.572 m}$

$h_8 = \sqrt{1.411^2 - (7 \times 0.16875)^2} - 0.411 = \textbf{0.361 m}$

$h_9 = \textbf{0.0 m}$

5.

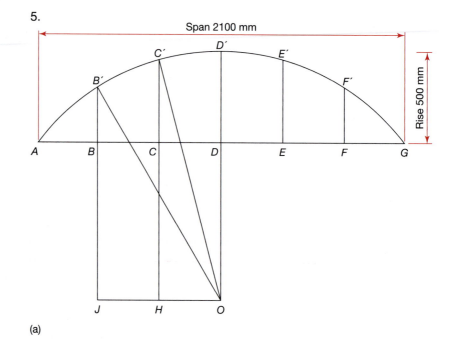

(a)

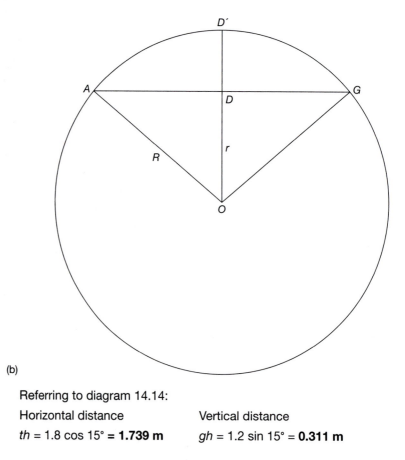

(b)

Referring to diagram 14.14:

Horizontal distance

$th = 1.8 \cos 15° = \textbf{1.739 m}$

Vertical distance

$gh = 1.2 \sin 15° = \textbf{0.311 m}$

$tk = 1.8 \cos 30° = $ **1.559 m** $jk = 1.2 \sin 30° = $ **0.600 m**

$tm = 1.8 \cos 45° = $ **1.273 m** $lm = 1.2 \sin 45° = $ **0.849 m**

$tp = 1.8 \cos 60° = $ **0.900 m** $np = 1.2 \sin 60° = $ **1.039 m**

$tr = 1.8 \cos 75° = $ **0.466 m** $qr = 1.2 \sin 75° = $ **1.159 m**

6. Using a similar diagram and labelling to that in the previous question:

$th = 1.6 \cos 15° = $ **1.545 m** $gh = 1.4 \sin 15° = $ **0.362 m**

$tk = 1.6 \cos 30° = $ **1.386 m** $jk = 1.4 \sin 30° = $ **0.700 m**

$tm = 1.6 \cos 45° = $ **1.131 m** $lm = 1.4 \sin 45° = $ **0.990 m**

$tp = 1.6 \cos 60° = $ **0.800 m** $np = 1.4 \sin 60° = $ **1.212 m**

$tr = 1.6 \cos 75° = $ **0.414 m** $qr = 1.4 \sin 75° = $ **1.352 m**

Exercise 15.1

1. Area of foundation = $2[(8.6 + 0.45 + 0.45) \times 0.45 + (7.1 \times 0.45)]$

     $= 2[(9.5) \times 0.45 + (3.195)] = 14.94 \text{ m}^2$

Volume of concrete = $14.94 \times$ thickness = $14.94 \times 0.750 = 11.205 \text{ m}^3$

For preparing 11.205 m^3 of concrete the quantities of materials and their costs are:

Material	Mass	
Cement	$240 \times 11.205 = 2689$ kg	108×25 kg bags
Fine aggregates	$720 \times 11.205 = 8068$ kg	10×850 kg jumbo bags
Coarse aggregates	$1440 \times 11.205 = 16\ 135$ kg	19×850 kg jumbo bags

Cost of cement = $108 \times £4.90 = £529.20$

Cost of fine aggregates = $10 \times £36.00 = £360.00$

Cost of coarse aggregates = $19 \times £36.00 = £684.00$

Total cost of materials = **£1573.20**

Labour hours = $2 \times 11.205 = 22.41$

Labour cost = $22.41 \times £15.00 = $ **£336.15**

2. Area of the wall = $5.5 \times 2.7 - 2(1.8 \times 1.2)$

     $= 14.85 - 4.32 = 10.53 \text{ m}^2$

Number of bricks = $10.53 \times 60 = 631.8$

Number of bricks with 5% extra allowance = $631.8 \times 1.05 = 663.39$, so **664**

Number of blocks = $10.53 \times 10 = 105.3$

Number of blocks with 5% extra allowance = $105.3 \times 1.05 = 110.57$, so **111**

Mortar required for brickwork = $0.026 \times 10.53 = 0.2738 \text{ m}^3$

Quantity of mortar with 10% extra allowance = 1.10 × 0.2738
$$= 0.3012 \text{ m}^3$$

Mortar required for blockwork = 0.012 × 10.53 = 0.1264 m³

Quantity of mortar with 10% extra allowance = 1.10 × 0.1264
$$= 0.139 \text{ m}^3$$

Brickwork

Assume the density of 1:3 cement/sand mortar to be 2300 kg/m³

Mass of 0.3012 m³ of mortar = 0.3012 × 2300 = 692.76 kg

Mass of cement = $\frac{1}{4}$× 692.76 = 173.19 kg

Mass of sand = $\frac{3}{4}$× 692.76 = 519.57 kg

Blockwork

Assume the density of 1:6 cement/sand mortar to be 2300 kg/m³

Mass of 0.139 m³ of mortar = 0.139 × 2300 = 319.7 kg

Mass of cement = $\frac{1}{7}$× 319.7 = 45.67 kg

Mass of sand = $\frac{6}{7}$× 319.7 = 274.03 kg

Total quantity of cement = 173.19 + 45.67 = **218.86 kg**

Total quantity of sand = 519.57 + 274.03 = **793.6 kg**

Cost of labour

Brickwork = £36.00 × 10.53 = **£379.08**

Blockwork = £9.00 × 10.53 = **£94.77**

3. (a) Area of the floor = 5.0 × 4.2 = 21.0 m²

Number of packs of floorboards = $\dfrac{21.0}{\text{coverage provided by one pack}}$

$$= \frac{21.0}{1.71} = 12.28$$

Wastage of 10% = 12.28 × $\dfrac{10}{100}$ = 1.228

Total number of packs required = 12.28 + 1.228 = 13.51, so 14

Cost of material = 14 × £29.50 = **£413.00**

Labour cost = 21.0 × £10.00 = **£210.00**

Total cost = £413.00 + £210.00 = **£623.00**

(b) Area of the room = 4.5 × 3.9 = 17.55 m²

Area of one sheet of chipboard = 2.4 × 0.6 = 1.44 m²

Number of sheets required = $\dfrac{17.55}{1.44}$ = 12.19

Wastage of 10% = 12.19 × $\dfrac{10}{100}$ = 1.22

Total number of chipboard sheets = 12.19 + 1.22 = 13.41 or 14

Cost of material = 14 × £11.50 = **£161.00**

Labour cost = 17.55 × £6.00 = **£105.30**

Total cost = £161.00 + £105.300 = **£266.30**

4. Area of roughcast finish on the wall = 2 × 9.15 ×2.95 + 2 × 7.0 × 2.95 − 5 ×(1.5 × 1.3) − 2.0 × 0.940 − 2.0 × 1.3 − 1.2 × 1.3

= 79.495 m^2

Volume of paint required = $\dfrac{\text{area of wall}}{\text{spreading rate of paint}} = \dfrac{79.495}{3}$ = 26.5 litres

Two coats of paint require 2 × 26.5 or 53 litres of paint

Number of five-litre cans = $\dfrac{53}{5}$ = 10.6 or 11

Cost of paint = 11 × £25.39 = **£279.29**

Labour cost (two coats) = 79.495 × £5.80 = **£461.07**

Exercise 16.1

1. Mean =1/12(5.5 + 6.0 + 6.0 + 6.3 + 6.5 + 7.1 + 7.2 + 7.7 + 7.8 + 8.0 + 8.2 + 8.5)

   Mean = **7.07%**

   Mode = **6.0%**

   Place in rank order: 5.5, 6.0, 6.0, 6.3, 6.5, 7.1, 7.2, 7.7, 7.8, 8.0, 8.2, 8.5

   Median = (7.1 + 7.2)/2 = **7.15%**

2. Mean = (11.5 + 14 + 11 + 12 +13.5 +10 + 12.5 + 12.8 + 13 + 13.3 + 11.4 + 13.6 + 15 + 10.5 + 9.8)/15 = 183.9/15 = **12.26%**

   There is no mode.

   Place in rank order: 9.8 + 10 + 10.5 + 11 + 11.4 + 11.5 + 12+ 12.5 + 12.8+ 13 + 13.3 +13.5 + 13.6 + 14 + 15

   The median = the eighth term = **12.5%**

   The range is 15 − 9.8 = **5.2%**

3. (a)

Class interval (crushing strength: N/mm^2)	Class midpoint (x)	Frequency (f)	f × x
32–34	33	3	99
35–37	36	7	252
38–40	39	8	312
41–43	42	7	294
44–46	45	9	405

47–49	48	6	288
50–52	51	5	255
		$\sum f = 45$	$\sum fx = 1905$

(b) Mean = $\dfrac{\sum fx}{\sum f}$ = 1905/45 **= 42.33 N/mm²**

4.

(a)

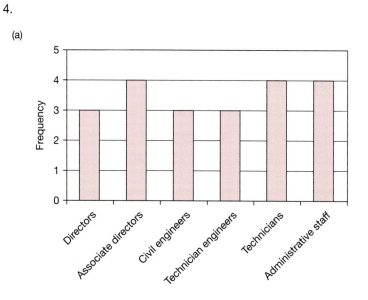

(b)

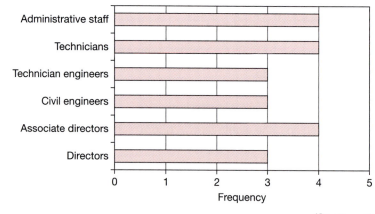

(Continued)

(c)

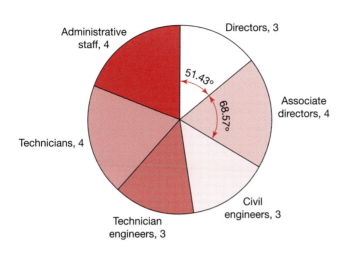

5.

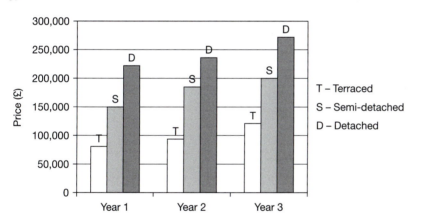

6.

(a)

Accidents in trenches/excavations

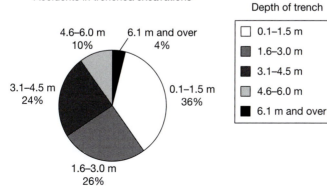

(Continued)

(b)

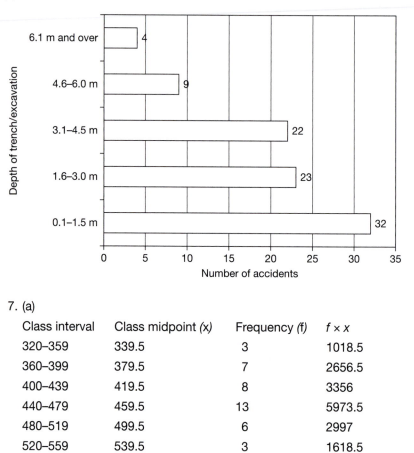

7. (a)

Class interval	Class midpoint (x)	Frequency (f)	f × x
320–359	339.5	3	1018.5
360–399	379.5	7	2656.5
400–439	419.5	8	3356
440–479	459.5	13	5973.5
480–519	499.5	6	2997
520–559	539.5	3	1618.5
		$\sum f = 40$	$\sum fx = 17\ 620$

(b)

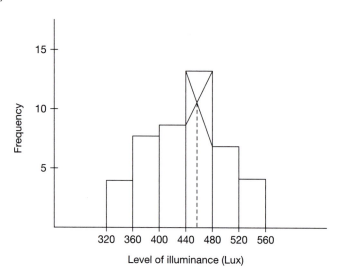

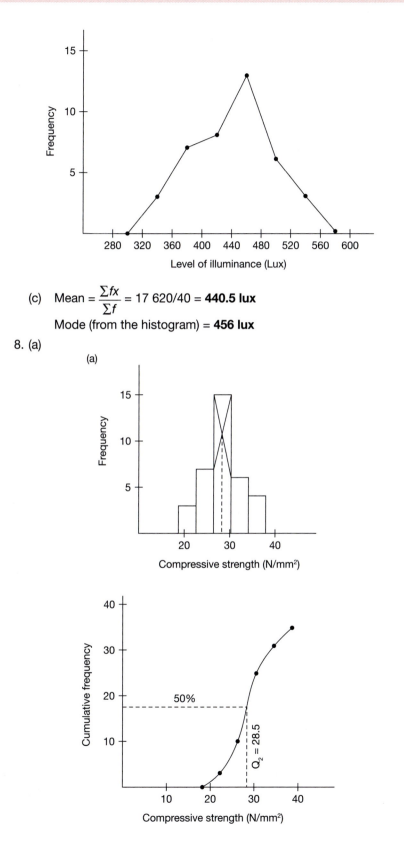

(c)   Mean = $\dfrac{\Sigma fx}{\Sigma f}$ = 17 620/40 = **440.5 lux**

Mode (from the histogram) = **456 lux**

8. (a)

Class interval (comp. strength)	Class midpoint x	Frequency f	$f \times x$	Comp. strength less than	Cumulative frequency
				18.5	0
19–22	20.5	3	61.5	22.5	3
23–26	24.5	7	171.5	26.5	10
27–30	28.5	15	427.5	30.5	25
31–34	32.5	6	195	34.5	31
35–38	36.5	4	146	38.5	35
		$\sum f = 35$			$\sum fx = 1001.5$

(b) Mean $= \dfrac{\sum fx}{\sum f} = 1001.5/35 = $ **28.6 N/mm²**

Mode = **28.5 N/mm²** (from the histogram)

Median = **28.5 N/mm²** (from the cumulative frequency curve)

9. (a)

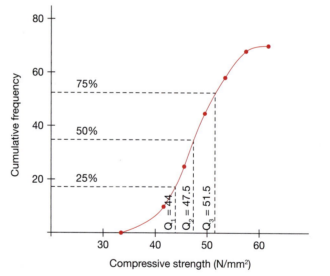

Class interval (comp. strength)	Frequency	Comp. strength – less than	Cumulative frequency
		33.5	0
34–37	2	37.5	2
38–41	8	41.5	10
42–45	15	45.5	25
46–49	20	49.5	45
50–53	13	53.5	58
54–57	10	57.5	68
58–61	2	61.5	70

(b) The median is the 50th percentile on the graph = **47.5 N/mm²**

  Inter-quartile range = 51.5 N/mm² – 44 N/mm² = **7.5 N/mm²**

Exercise 17.1

1. Slant height from centre of base to apex = $\sqrt{\dfrac{35^2}{2^2} + 7.5^2}$ = 19.0394 m

  Surface area of one side = ½(35 × 19.0394) = 333.19 m²

  Surface area of four sides = 4 × 333.1895 = **1332.76 m²**

2. Slant height, $l = \sqrt{60^2 + 25^2}$ = 65 m

  Surface area = $\pi r l = \pi \times 25 \times 65$ = **5105.09 m²**

3. Area = 5105.09 – 500 = **4605.09 m²**

4. (a) Slant height = $\sqrt{20^2 + 6^2}$ = 20.8806 m²

  Area of one side = $\left(\dfrac{45 + 5}{2}\right) \times 20.8806$ = 522.0153 m²

  Area of four sides + top = 4 × 522.0153 + 25 = **2113.06 m²**

  (b) Volume = $\dfrac{1}{3} \times 6 \left(45 \times 45 + \sqrt{(45 \times 45) \times (5 \times 5)} + 5 \times 5\right)$

  $= \dfrac{1}{3} \times 6 \,(2025 + 225 + 25)$ = **4550 m³**

5. Volume of top part = $\dfrac{1}{3}\pi h\,(r^2 + rR + R^2)$

  $= \dfrac{1}{3}\pi \times 4.5\,(1^2 + 1 \times 2 + 2^2)$ = 32.987 m³

  Volume of right cone = $\dfrac{1}{3}\pi r^2 h = \dfrac{1}{3}\pi \times 1^2 \times 0.5$ = 0.524 m³

  Total volume = 32.987 + 0.524 = **33.51 m³**

6. Slant height, $l = \sqrt{1^2 + 4.5^2}$ = 4.61 m

  Slant height of small cone, $l_1 = \sqrt{1^2 + 0.5^2}$ = 1.118 m

  Surface area of both vessels = $\pi l(r + R) + \pi r l_1$

  $= \pi \times 4.61(1 + 2) + \pi \times 1 \times 1.118$

  $= 43.448 + 3.512$ = **46.96 m²**

7. Let d = diameter of smaller end of frustum

  (a) then $\dfrac{d}{150} = \dfrac{150}{350}$ ∴ d = **64.29 mm**

  (b) Slant height of frustum = $\sqrt{200^2 + 42.857^2}$ = 204.54 mm

  Slant height of cone = $\sqrt{75^2 + 350^2}$ = 357.946 mm

  Surface area of cone = $\pi \times 75 \times 357.946$ = 84 338.93 mm²

  Surface area of frustum:

  $= \pi \times 204.54(75 + 32.1429) + \pi \times 32.1429^2$ = 72 093.82 mm²

Difference in area = 84338.93 – 72093.82 = 12 245.11 mm²

Therefore the cone is larger by **12 245.11 mm²**

Exercise 18.1

1. Mid-ordinates are: 0.5(1 + 1.5) = 1.25 cm

   0.5(1.5 + 3) = 2.25 cm

   0.5(3 + 5.5) = 4.25 cm

   0.5(5.5 + 9) = 7.25 cm

   0.5(9 + 13.5) = 11.25 cm

   Area by mid-ordinate rule = 1(1.25 + 2.25 + 4.25 + 7.25 + 11.25)

   = 26.25 cm²

   Area by trapezoidal rule = 1{0.5(1 + 13.5) + 1.5 + 3 + 5.5 + 9}

   = 26.25 cm²

   Both results are higher than the exact answer.

2. By trapezoidal rule:

   15{0.5(45 + 43) + 52 + 56 + 53 + 49 + 45} **= 4485 m²**

   By mid-ordinate rule area equals **4485 m²**

   By Simpson's rule, area = 15/3{45 + 43 + 2(56 + 49) + 4(52 + 53 + 45)

   = 4490 m²

3. Area by Simpson's rule = 3/3{2 + 10.9 + 2(3.3 + 6.3) + 4(2.5 + 4.6 + 8.2)}

   = 93.3 cm²

4. (a) Area by trapezoidal rule = w[½(first + last) + sum of remaining ordinates]

   = 12[½(1.25 + 2.15) + 1.25 + 1.4 + 1.65 + 1.6 + 1.6 + 1.8 + 1.9]

   = 12[(1.7) + 11.2] = **154.8 m²**

   (b) Volume = 154.8 × 1.2 = **185.76 m³**

5. Area by trapezoidal rule = w[½(first + last) + sum of remaining ordinates]

   = 12[½(0 + 0) + 23 + 46 + 63 + 66 + 51 + 28]

   = 12[277] = **3324 m²**

6. Your answer depends on the accuracy of your draughtsmanship. But here a different method is given to show accurate answers.

   In $\triangle ACB$, $\dfrac{114}{\sin 50} = \dfrac{100}{\sin A}$, $\therefore \angle A = 42.219°$

   $\therefore \angle B = 180 - 50 - 42.219 = 87.781°$

   Area of $\triangle ACB$ = 0.5(114)(100) sin 87.781 = 5695.726 m²

   In $\triangle ACB$, $\dfrac{114}{\sin 50} = \dfrac{b}{\sin 87.78}$, $\therefore b = 148.705$ m

   In $\triangle DCA$, $148.705^2 = 95^2 + 105^2 - 2(105)(95) \cos D$

   $\therefore \angle D = 95.936°$

Area of ΔDCA = 0.5(105)(95) sin 95.936 = 4960.757 m²

Total area = 5695.726 + 4960.757 = **10 656.48 m²**

7. Area = $\frac{1}{3}$ w [(first + last) + 4 (sum of the even ordinates) + 2 (sum of the remaining odd ordinates)]

$= \frac{1}{3} \times 20[(0 + 0) + 4(1.2 + 2.2 + 2.6 + 0.7) + 2(1.7 + 3.0 + 1.8)]$

$= \frac{1}{3} \times 20[(0 + 0) + 4(6.7) + 2(6.5)] = 265.333$ m²

Therefore volume = area × width of excavation

= 265.333 × 20 = **5306.67 m³**

8. Volume = $\frac{1}{3} \times 10[3.5 + 2.6 + 2(3.0) + 4(3.3 + 2.75)] =$ **121 m³**

9. Area of top = 55 × 49 = 2695 m²

Area base = 48 × 43 = 2064 m²

Area mid-section = ½(49 + 43) × ½(55 + 48) = 2369 m²

Volume = $\frac{3.3}{6}$ [2695 + 4(2369) + 2064] = **7829.25 m³**

10. The cross-sections are triangles.

Area of left-hand side = ½(15 × 12) = 90 m²

Area of right-hand side = ½(40 × 17) = 340 m²

Area mid-section = ½[½(40 + 15) × ½(17 + 12)] = 199.375 m²

Therefore volume = $\frac{30}{6} \times [90 + 4(199.375) + 340] =$ **6137.5 m³**

Exercise 19.1

1. If side a = side c then $\angle A = \angle C =$ **72°**

$\angle B = 180 - 2(72°) =$ **36°**

$\frac{b}{\sin 36} = \frac{50}{\sin 72}$

Side $b = \frac{50 \times \sin 36°}{\sin 72°} =$ **30.9 cm**

2. $\angle A = 180° - 40° - 55° =$ **85°**

$\frac{30}{\sin 85°} = \frac{c}{\sin 55°} = \frac{b}{\sin 40°}$

$\therefore c = \frac{30 \times \sin 55°}{\sin 85°} =$ **24.67 cm**

and $b = \frac{30 \times \sin 40°}{\sin 85°} =$ **19.36 cm**

3. In ΔEGC, CE = 2.666 cos 30° = **2.309 m**

GE = 2.666 sin 30° = **1.333 m**

In ΔAGC, $\angle A = \angle C$

Therefore side AE = side CE = **2.309 m**

Also, side AG = side CG = **2.666 m**

$\triangle AFG$ is an equilateral triangle, $\therefore$ $FG = AG$ = **2.666 m**

4. $q^2 = 25^2 + 30^2 - 2(25)(30) \times \cos 44°$

   $q^2 = 445.99$

   $\therefore q$ = **21.12 cm**

   $\dfrac{30}{\sin P} = \dfrac{21.1184}{\sin 44°}$

   $\therefore \angle P$ = **80.681°**

   and $\angle R = 180 - 44 - 80.681$ = **55.319°**

5. $a^2 = b^2 + c^2 - 2bc \cos A$

   $8^2 = 6^2 + 10^2 - (2 \times 6 \times 10) \cos A$

   $\cos A = 0.6$, $\therefore$ $\angle A = \cos^{-1} 0.6$ = **53.13°**

   $b^2 = c^2 + a^2 - 2ca \cos B$

   $6^2 = 10^2 + 8^2 - (2 \times 10 \times 8) \cos B$

   $\cos B = 0.8$, $\therefore$ $\angle B = \cos^{-1} 0.8$ = **36.87°**

   $\angle C = 180 - 53.13° - 36.87°$ = **90°**

6. Area = $\frac{1}{2}(200)303 \times \sin 46° = 21\,795.996$ m^2

   Area = $\frac{1}{2}(303)192 \times \sin 48° = 21\,616.597$ m^2

   Total area = **43 412.59 m^2**

7. In $\triangle BDC$, $\dfrac{114}{\sin 50°} = \dfrac{100}{\sin B}$, $\therefore \angle B = 42.219°$

   $\angle C = 180° - 50° - 42.219°$ = **87.781°**

   $\dfrac{c}{\sin 87.781°} = \dfrac{114}{\sin 50°}$, $\therefore c = 148.705$ m

   In $\triangle ADB$, $148.705^2 = 95^2 + 105^2 - 2(95)(105) \times \cos A$

   $\cos A = -0.103417$, or $\angle A$ = **95.936°**

   $95^2 = 148.705^2 + 105^2 - 2(148.705)(105) \times \cos B$

   $\angle B = 39.4514°$

   $\angle D = 180° - 95.936° - 39.4514° = 44.6126°$

   $\therefore \angle ADC = 50° + 44.6126°$ = **94.6126°**

   $\therefore$ In $\triangle BDC$, $\angle B = 180 - 50 - 87.781 = 42.219°$

   $\therefore \angle ABC = 42.219 + 180 - 44.614 - 95.936$

   $\qquad\qquad$ = **81.671°**

8. In $\triangle ADC$, $\angle A = 180° - 95° - 41° = 44°$

   $\dfrac{500}{\sin 44°} = \dfrac{AD}{\sin 41°}$, $\therefore$ AD = **472.217 m**

   $\dfrac{AC}{\sin 95°} = \dfrac{500}{\sin 44°}$, $\therefore AC = 717.039$ m

   In $\triangle BDC$, $\angle B = 180° - 43° - 88° = 49°$

$\dfrac{500}{\sin 49°} = \dfrac{BC}{\sin 43°}$, ∴ BC = **451.828 m**

In △ABC, ∠C = 88° − 41° = 47°

$AB^2 = 717.039^2 + 451.828^2 - 2(717.039)(451.828) \times \cos 47°$

$AB^2 = 276\,388.134$, ∴ AB = **525.726 m**

9. In △EHG, $\dfrac{EH}{\sin 38°} = \dfrac{500}{\sin 40°}$, ∴EH = **478.899 m**

In △FHG, $\dfrac{FG}{\sin 51°} = \dfrac{500}{\sin 49°}$, ∴FG = **514.864 m**

$\dfrac{FH}{\sin 80°} = \dfrac{500}{\sin 49°}$, ∴FH = 652.442 m

In △EHF, $EF^2 = 652.442^2 + 478.899^2 - 2(652.442)(478.899) \times \cos 51°$

$EF^2 = 261\,757.6935$, ∴EF = **511.623 m**

In △EHG, $EG^2 = 500^2 + 478.899^2 - 2(500)(478.899) \times \cos 102°$

∴EG = 760.864 m

In △EFG, $\dfrac{511.623}{\sin 42°} = \dfrac{514.864}{\sin E}$

sin E = 0.67337, ∴∠E = 42.328°

∠HEF = 40° + 42.328° = **82.328°**

∠EFG = 360° − (82.328° + 102° + 80°) = **95.672°**

Exercise 20.1

1.

	Length (m)	Width (m)	Area (m²)	Number of packs	Wastage @ 10%	Number of packs required
DINING	3.5	3.5	=B4*C4	=D4/2.106	=10/100*E4	=E4+F4
LIVING	6.1	4.2	=B6*C6	=D6/2.106	=10/100*E6	=E6+F6
HALL	4	2	=B8*C8	=D8/2.106	=10/100*E8	=E8+F8
TOTAL						=SUM(G4:G8)

2.

Microsoft Excel - Exercise 2(ch20).xls

	A	B	C	D	E	F
1						
2		LENGTH (m)	WIDTH (m)	DEPTH (m)	VOLUME (m^3)	
3						
4	A	7.5	0.5	0.75	=B4*C4*D4	
5	B	3.5	0.5	0.75	=B5*C5*D5	
6	C	3.5	0.4	0.75	=B6*C6*D6	
7	D	3.5	0.5	0.75	=B7*C7*D7	
8						
9				TOTAL	=SUM(E4:E7)	

3.

Microsoft Excel - Exercise 3(ch20).xls

	A	B	C	D	E	F	G	H	I	J	K	L	M	N	O	P	Q
1																	
2					Maximum Tensile Force (kN)											Mean	Range
3	9.9	12.1	10	12	10.2	11.9	10.3	11.6	10.5	11.5	11.3	10.9	11.3	11.6	11	=AVERAGE(A3:O3)	=MAX(A3:O3)-MIN(A3:O3)

4.

	A	B	C	D	E	F	G
1							
2	class interval	frequency (f)	class mid-point (x)	fx	Mean (Σfx/Σf)		
3							
4	60 - 64	2	62	=B4*C4			
5	65 - 69	6	67	=B5*C5			
6	70 - 74	11	72	=B6*C6			
7	75 - 79	5	77	=B7*C7			
8	80 - 84	1	82	=B8*C8			
9							
10		Σf =		Σfx =			
11		=SUM(B4:B8)		=SUM(D4:D8)	=D11/B11		
12							

5.

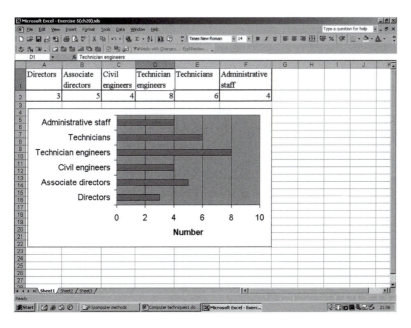

	A	B	C	D	E	F
1	Directors	Associate directors	Civil engineers	Technician engineers	Technicians	Administrative staff
2	3	5	4	8	6	4

6.

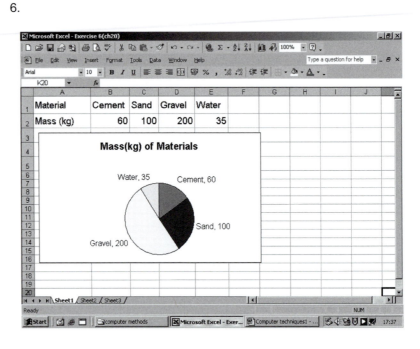

7.

Component	U-value (W/m²)	Area (m²)	Temperature difference	Heat Loss (Watts) = U x A x T
Walls	0.35	47.3	20	=B3*C3*D3
Floor	0.35	20	20	=B4*C4*D4
Roof	0.3	20	20	=B5*C5*D5
Patio door	2.6	5	20	=B6*C6*D6
Door	0.46	1.7	20	=B7*C7*D7
Total				=SUM(E3:E7)

Index